Earthways

Earthways

Simple Environmental Activities
For Young Children

By Carol Petrash
Illustrations by Donald Cook

gryphon house

Mt. Rainier, Maryland

© 1992 Carol Petrash
Published by Gryphon House, Inc.
3706 Otis Street, Mt. Rainier, MD 20712

Library of Congress Catalog Number: 92–53892

⊕ Printed in the United States of America on recycled paper using soy-based inks.

Design: Graves, Fowler & Associates
Cover Illustration: Copyright © 1992 Scott Gustafson

Publisher's Cataloging in Publication
(Prepared by Quality Books, Inc.)

Petrash, Carol A., 1947-
     Earthways : simple environmental activities for young children / Carol Petrash.
     p. cm.
     Includes bibliographical references and index.
     ISBN 0-87659-156-X
     1. Environmental education—Activity programs. 2. Environment and children. 3. Conservation study and teaching. 4. Education, Preschool—Activity programs. I. Title. II. Title: Simple environmental activities for young children.

GF26.P4 1992   375'.008'3
                                                              QBI92-802

For Jonathan, Joshua and Ava,

my children,

and

For All Children

*May joyful experience and warm feelings mature
into imaginative thoughts, and may your hands work always in
harmony with the ways of the Earth.*

"What is honored in a country will be cultivated there."

—Plato

Many thanks to

. . . Jack, my husband, for his endless enthusiasm, unwavering confidence and countless hours at the keyboard.

. . . Kathy, my friend and editor, for her clear thoughts and kind ways.

. . . Joan, mentor and friend, for suggestions on the manuscript and practical advice.

. . . Chip and Leah for believing, for a long time, that I had it in me.

. . . Sarabeth and all the people at Gryphon House who helped make this happen.

. . . Don Cook, for saying yes and illuminating this book with his drawings.

. . . And lastly, I am deeply indebted to the work of Rudolf Steiner and to those who have carried the health-bringing ideals and principles of Waldorf Education to children and teachers all over the world. My work in Waldorf kindergartens has made it possible for me to write this book.

Chapter 4 Summer

The Whole Earth Home and Classroom

Bringing Nature In

Resources

Introduction

The environmental problems that confront us today are as varied as they are severe. Global warming, the depletion of the ozone layer, the destruction of the rain forests are large problems that seem at times insurmountable. At the same time, in the last twenty years, environmental awareness has steadily increased. Clearly, the future of our planet will depend on the degree to which all our children are made environmentally aware, respectful of our planet Earth and its natural kingdoms, and sensitive to a higher quality of Nature.

The task is large, yet children are naturally suited for this challenge. They come into life with a sense that the world is good and beautiful. Our interactions with them and the ways in which we bring them into contact with nature can either enhance these intuitions or destroy them. When children are met with love and respect, they will have love and respect to give. Our task as the parents and educators of young children is not to make them frightfully aware of environmental dangers, but rather to provide them with opportunities to experience what Rachel Carson called "the sense of wonder." Out of this wonder can grow a feeling of kinship with the Earth.

We must allow them to play with the elements of earth, air, water and, carefully supervised, to experience the quiet power and beauty of fire. Through nature crafts and natural toys and opportunities to joyfully experience and celebrate the seasons of the year with all that nature brings, we can hope to plant the seeds of a new attitude of reverence toward nature. This childhood experience may then mature, when they are adults, into a love-warmed thinking which will not allow them to treat the Earth as a possession or commodity that they have a right to exploit, but rather as a precious gift which is their honor and duty to protect and enhance.

Joseph Cornell, environmental educator and author of *Sharing Nature With Children,* stated this so well when he said that children need to build a strong foundation of love for the environment so that reason will follow feeling.

We will serve our children well when we turn their attention to the stones, the grasses and flowers, the trees and animals by finding meaningful and creative ways to allow nature to enter our classrooms and play areas, and by bringing our children out into nature as much as possible. A loving relationship with nature will not only promote health for our planet but health for our children as well. Contact with nature can bring simplicity back into the lives of children (and parents and teachers!): the simplicity of nature's pace, the simplicity of the seasons. All children should have the opportunity to delight in the simple sound of a bird's song; in the texture of a tree's bark; in the sweet smell of freshly turned earth.

In addition, we must find gentle and loving ways to show the children sound environmental practices such as recycling, composting and, above all, thrift. For young children, leading by example is preferable to heavy-handed explanations which leave a residue of guilt and fear. Environmental awareness will come naturally to children when it is integrated into the early childhood classroom and home as a way of life.

How To Use This Book

This book is organized seasonally, that is, there are large sections for fall, winter, spring and summer. Each of these seasonal sections has certain subsections. They are The Whole Earth Home and Classroom; Bringing Nature In: The Season's Garden; Bringing Nature In: Seasonal Crafts; and Supplying the Missing Links.

The Whole Earth Home and Classroom section contains material which can help you with the "greening" of your classroom, for example, help you make your classroom or home more environmentally friendly. The seasonal focus for each of these sections is as follows:

Fall—Setting Up An Earth-Friendly Classroom: Breaking the Throwaway Habit

Winter—Toys From Nature: The Indoor Play Space

Spring—Cleaning House: Using Earth-Friendly Products and Materials

Summer—Creating a More Natural Outdoor Play Space

In *Bringing Nature In: The Season's Garden*, you will find information on how to set up a Seasonal Garden indoors which can help you celebrate the unique qualities of each season and establish an important connection to the rhythm of the passing of time throughout the year.

Bringing Nature In: Seasonal Crafts may surprise you in that the crafts do not all focus on using materials from nature. For example, the winter section contains sewing activities. While I recommend that you use wool felt, a natural material, for the sewing projects, the connection to the season is through the activity—a more quiet, indoor, focused one appropriate for the wintertime when nature (and people) manifest a more quiet mood. The key is to connect with nature and the rhythm of the year not just through the use of natural materials, but by recognizing the special character or mood of each season and expressing that through appropriate activities. We all do this already to a certain extent. For example, we might not blow bubbles outdoors in wintertime, but we would save it for a breezy spring or summer day. The trick is to become more conscious of this idea and capitalize on it in our planning year round.

This might mean taking a look at how you organize project time. Many of the projects described in this book can be done for several consecutive days or even a week or more. In fact, it benefits the children to be able to experience something in this way. With the advent of television generations, we tend to feel that children need lots of change and stimulation. We worry that repetition might bore them. Don't accept this attitude! When we allow children to deepen their experience of anything they do by repetition, the sense of mastery and familiarity that results protects and strengthens their developing sense of self. This gift of time is, for young children especially, a precious one.

If you can structure your class time so that the projects can take place during the free play time rather than at a special project time when everyone must participate at the same time, it will allow both teachers and children the time and space to really get involved. Set up a working space—a sewing corner, a baking table, whatever—and get started while the children are engaged in their play. The power of the activity and their natural curiosity and inclination to imitate will draw them to you, though probably not all at the same time. Some will be very engaged in their own work (their play). If you want all the children to participate in a particular activity, keep track of who doesn't come and then bring them to it at some point. This can often be accomplished imaginatively by entering into the mood of their play. For example, the shopkeepers need to come help make some toys for their shop, or the firefighters want to knead some bread for their snack. This allows the children to fit the new activity into their stream of play rather than feel it as an abrupt change of direction or an interruption. Then, too, there may be times when a child is not able to find his or her niche during play and bringing them to the more focused, teacher-directed craft activity will be just what they need. They will often find renewed direction of their own after a time of purposeful work such as this. This approach lends itself to activities that take several days to complete, giving all the children a chance to take part. Try, then, to expand your idea of nature crafts to include not only natural materials, but also "natural time," for example, finding the appropriate activity for a particular time of year and the appropriate way in which to bring it to the children. These approaches are appropriate not only for the classroom, but at home as well.

Supplying the Missing Links provides activities that will allow the children to connect a product which they often use and usually purchase in a store with the source and process from which it comes. The aim is that they will then have a subtle understanding of their strong connections with and dependence on the Earth and an experience of making things for themselves. For example, they will see that bread comes from wheat and experience the process of how this occurs.

They will also have the experience of using the whole of something. So, for example, the Halloween jack-o'-lantern can later be cooked for pies and its seeds roasted and saved for the birds and squirrels while the remaining innards go to the compost pile and, eventually, back into the garden to enrich the soil. The whole is used. This was a principle of life for the Native Americans and one that is supremely ecological and environmentally sound. Children are intuitively aware of the wholeness of life. Our task is to confirm this connection for them by rededicating ourselves to this ideal.

Lastly, they will have the experience of "I can do it and do it well!" They will make their own cards and gifts—things that are useful as well as beautiful—the product of their hands and hearts. They can begin to become creators rather than consumers.

Chapter 1
Fall

The Whole Earth Home and Classroom

Setting Up an Earth-Friendly Home and Classroom: Breaking the Throwaway Habit

We have all grown up with the notion that more is better; that if something breaks, we replace it with something new; that there is an endless supply of whatever we need. These ideas are no longer viable, and we need to raise our children with more earth-aware habits and attitudes.

How can homes and classrooms be more earth-friendly? Start by using fewer and fewer items that are thrown away—for example, paper napkins, paper towels and plastic cutlery. First check with your local health department or licensing agency about their requirements and explain to them what you are trying to do. Then, if possible, start using small pottery cups instead of disposable paper cups. Mark each child's name on the bottom of his or her cup—so the children always use the same cup. After each use wash the cups and let them air dry. A local potter made us a set of small mugs, but you could also have each child bring a nice small mug from home.

Another earth-friendly idea is to make small cotton placemats (approximately 8 inches by 6 inches) for the children to use instead of paper napkins. Gradually (with the help of parents and friends, yard sales and thrift shops) accumulate a nice set of small bowls, spoons, forks and knives for the classroom. Children can learn earth-friendly habits by setting the table with the fabric placemats for snack. Just make sure there are the correct number of chairs at each table, and have the children put out the placemats and cups, or whatever is needed. (This is excellent practice in learning one-to-one correspondence!)

Set up a dish washing station near the sink with two tubs—one with hot, soapy water and a dish mop, the other with clear, warm water for rinsing—and a drying rack or a dish drain. When the children are excused from the table, they wash their dishes before going to the rug for rest or circle. Once this becomes a habit—and it is an excellent one—they become very good at it and actually need little supervision. If you have a class with only three year olds, you may want the assistant teacher to wash the dishes with the children's help.

Wash the placemats weekly as an activity on your regular wash day. Use two wash tubs again—one with soapy water and one with rinse water—and a small old-fashioned scrub board (still available in hardware stores) and scrub soap (also still available in grocery and hardware stores, although you may have to ask for it). Scrub soap is a hard bar soap that doesn't get mushy in water. Set the tubs on two thick bath towels to catch some of the water, but also just figure it is water play day. Begin by washing—wetting the mats, rubbing them on the bar of soap (it rests conveniently on the top edge of the scrub board) and scrub-a-dub-dubbing them up and down on the scrub board. Then into the rinse water, wring them out and hang them to dry. Use an indoor drying rack or place them in a basket to hang outdoors on a clothesline later. The children love to help with the washing, and, of course, you can always "volunteer" those who seem to need some watery work to focus or calm them. This is also a good time for washing doll's clothes or bedding or other classroom cloths or linens that need it. Once the washing is done, take out cups, ladles and egg beaters for water play.

Other Suggestions for Helping to Break the Throwaway Habit

Oil Cloth—Use oil cloth (often made of vinyl) instead of newspapers for covering tables during messy activities. It can be wiped clean and endlessly reused. It is more pleasant than newspaper to work on as the newsprint won't come off on your hands. Better to recycle your newspapers than have them wind up in the garbage can. Oil cloth is available at fabric, hardware or variety stores. It is also available mail order from the Vermont Country Store catalog—a good source for lots of "old-fashioned" things you may not be able to find at other places. (You will find the address and phone number in the Resources at the end of this book.)

Children's Cloth Aprons—These aprons can be washed and used over and over for cooking, baking, painting, etc. They can be easily made from thick kitchen or hand towels by using the pattern. Fold over the top corners (step one) and sew down to make

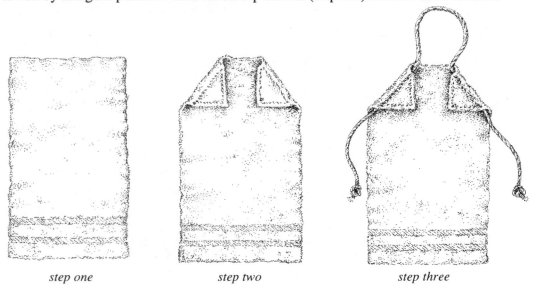

| step one | step two | step three |

two "pockets" (step two). Run a piece of thin rope or cord up through one pocket over the top and down through the other pocket (step three). Make big knots in the ends of the rope or cord so it will not easily slip back through the pockets. This cord slips over the head and ties behind the back to secure the apron.

Work Folders—Make a folder for each child by folding a large sheet of construction paper (approximately 12 inches by 16 inches) in half. Keep the children's drawings, paintings, etc., in these folders. When you have collected a large amount, make the children's work into a book, rather than letting it go home in dribs and drabs. When collected in this way, it can be made into an attractive and appreciated gift for parents. This helps children learn, from an early age, that their work is valued and should be treated accordingly. Encourage the children to use both sides of the paper when drawing.

Recycling—Most communities across the country are beginning to see the need for, and the good sense inherent in, recycling. Have a bin in your classroom (an extra trash can with a lid works well) in which to place items such as glass, plastic, metal and aluminum to be recycled. Remove the lids and rinse and air dry containers before placing them in the bin. Then just do it! If it becomes a habit for you, the children will imitate it. They don't need lectures about landfills filling up. Just tell them, in a matter-of-fact way, that it's good for the Earth when we save these things, so that they can be made into something new and used again! When the bin is full, take it to a recycling center. Enlist parents' help; perhaps children and families can take turns bringing materials to the recycling center or home for curbside pickup, if they are lucky enough to have it. Encourage your school to contract with a trash hauler who can arrange a recycling pick-up. It may actually save the school money, as the amount of actual trash you generate will decrease greatly, especially if the school is recycling paper as well.

Composting—Composting kitchen, garden and yard waste is a wonderful way to get excellent soil for gardening and planting projects. Check with your health department or local environmental agency about the requirements in your area. Generally speaking, if it grew out of the earth, you can compost it. The basic components of a compost pile are green things and brown things. This includes vegetable and fruit scraps, grass clippings, leaves, weeds, leftover scraps of bread from lunch, apple cores from snack preparation, etc. Do not include meats, fats, dairy products or packaging of any kind. You will need a small bucket or container for classroom collection of "compostables," preferably one with a lid. It is best to empty and rinse it daily.

The compost pile can be as simple as it sounds—a pile of things you want to compost standing in a quiet corner of the garden or playground. Or, you can buy a polystyrene or cedar composter from a garden center or hardware store. These have the advantages of being very tidy and animal proof, and they often meet health department requirements. They do cost some money; however, there are a number of reasonably priced ones on the market. You can find compost bins at garden centers, hardware stores or from several of the catalogs listed in the Resources section at the end of this book.

To make your own pile, dig a shallow hole about six inches deep and the size you want the pile to be (3 feet by 3 feet is a good size); put coarse materials in the bottom, such as small branches, sticks and weeds to allow air to circulate up through the pile; and then add the compost material as necessary. Keep a shovel nearby to chop up the things you add, and they will decompose more quickly. Cover the new additions with a thin layer of soil from the six inches that you excavated. You can also use grass clippings and sawdust as covering material. Leaf mulch (often available from local municipalities) and dried horse manure (available from local stables—even in New York City!) are also excellent for covering a pile. When your pile is well covered it is less likely to be disturbed by animals. Vary the kinds of materials you put in—again thinking of greens and browns. The little organisms that break down these materials can really get cooking if there is a variety. They also need a bit of moisture for optimum working conditions.

Again, composting can get much more fancy and detailed than this. This book is not the definitive work on composting; it is meant to show you that you **can** do it. It is possible to just let it rot! If you are in an urban area, one of the more "official" composters may be necessary to deter unwanted rodent visitors. It would be a bit of an investment, but one that would last for years. And if you tell the store or the catalog company who you are and what you are trying to do, they may respond in kind. Many companies of all kinds give discounts to schools. If you need more detailed information, try one of the books listed in the Resource section, or the local agricultural or extension agent in your area.

It is amazing to see the children's attitudes towards previously "yucky" stuff—pumpkin innards, leftover bread, etc.—change when it just, as "a matter of fact," goes into the compost bucket to be taken later to the compost pile to go back to the Earth. The children intuitively know about the rightness of these things, and it resonates in them when they see that the adults around them know and care as well.

Repairing Toys—When toys and equipment break, don't throw them away! Have a place, like a big basket, to collect things that are broken and need repair. Once every couple of weeks or so, take out tools, glue, toothpicks, etc. Bring the basket out, sit down and fix what was broken. You really will be able to fix most things, especially if you gradually replace plastic toys with those made of wood. The children will love to help. It gives them a wonderful sense of well-being to see things made whole again, and it helps them to develop the opposite of a throw-away consciousness. The children in my class used to believe that I could fix anything. They would even tell their parents that they could take broken items into class for repair—a wonderful, invaluable "can do" attitude!

Other Earth-Sound Ideas

• Buy supplies and materials in quantity to save on unnecessary packaging. When considering the packaging of products you buy, think about the source—plastic is made from a non-renewable resource, petroleum, whereas paper and cardboard are made from trees. Both types of packaging cause pollution during the production process. In terms of their recyclability, you will have to see what is possible in your area. Just because something is recyclable doesn't mean you will be able to recycle it in your community. Check and see, and base your product choices on materials that will have the least negative environmental impact. For further thoughts in this direction, see the spring Whole Earth Home and Classroom section.

• If you can find sensible alternatives to non-recyclable plastic, use them! For example, wax paper sandwich bags are a sound alternative to plastic ones. Cut-Rite® is now making them again. Ask for them at your grocery store.

• Look for child-sized cutting boards and knives (table or luncheon knives actually work and aren't dangerous) which allow the children to be involved in the preparation of their food. Using fresh fruits and vegetables for snack which the children can help prepare helps you avoid packaged snack foods that are neither environmentally sound nor always healthy for children to eat. Make snack time special by setting the table (a small vase of flowers adds a nice touch), settling everyone and setting the mood with a short verse of thanksgiving before you eat. I used the following:

> Earth who gave us all this food,
> Sun who made it ripe and good,
> Dearest Earth and dearest Sun
> We will not forget what you have done.
> —*Christian Morgenstern*

Bringing Nature In:
Creating a Seasonal Garden

Creating a small Season's Garden or corner in your home or classroom is one way to bring nature indoors and celebrate the rhythm of the seasons of the year. Young children thrive on rhythm—not the rigid holding to a timetable but the rhythmic flow of one thing into another. It gives them a sense of security and well-being to know that as it was, it shall be again. For many children today, a connection to nature and the passing of the seasons is one of the few constants in their lives. As such it is all the more important to emphasize this connection and provide a space for recognizing and celebrating it.

Where

Set aside a quiet corner somewhere in your classroom (or home) that is out of the way yet accessible to the children. It is nice to have a small table which can be the focus, but you could also use a shelf on a bookcase or an area on a counter top. A large tree stump (about 2–3 feet high) would be especially nice—keep your eyes open for tree work being done in your neighborhood. A small, round table is especially nice, as the roundness is a more natural shape and suggests the circle of the year.

What

Once you have decided where the Season's Garden will be, you can begin to create it. One of the first things to consider is color and how you will use it to create the appropriate seasonal mood. Think about the different seasons and which colors they suggest to you. This may be very individual and will certainly be related to the part of the world you're living in. In the mid-Atlantic area with four definite seasons, the following are suggested:

> Fall: Warm colors—soft reds, oranges and golden yellows
> Winter: Cool colors—deep blues, icy blues and violets
> Spring: Pastels—pinks, pale yellows, spring greens
> Summer: Fiery colors—deep intense reds, yellows, oranges

Find or dye small or medium-sized cloths in these various colors and keep them in a special box so you have them when needed. Thrift shops and flea markets are good sources for nice old linens, napkins, etc., that you can use. Dyeing is fun to do with the children. Natural fabric dyes are available from Earth Guild. (See Resource section at the end of this book.) If your Season's Garden focal point is a table, you may want to cover it with a white or other neutral color cloth and use smaller, seasonal cloths as color accents.

Another nice touch you can add is a medium blue backdrop cloth to represent the sky. This works especially well if you are setting up your Season's Garden in a corner. Hang the cloth from a point two to three feet above the garden, and drape it gently down and around the edges of the garden, tucking its edges under the seasonal cloths. Light or medium weight cheesecloth works very nicely for the sky cloth. Cheesecloth is available at fabric stores and mail order from Strauss and Company. (See the Resource section at the end of this book.) It sells in wholesale amounts and will send you a sample card of the various weights of cheesecloth on request. Actually, cheesecloth—particularly the medium weight—is a wonderful, fairly inexpensive material to have in the home or classroom for cloths of all kinds, from those you use in the Season's Garden to play cloths for the children. Get some friends or other teachers together and buy it by the box. It's even cheaper that way.

Now What

Your Season's Garden now has a home and a colorful beginning. What else does it need? Remember that this little place is one in which to celebrate the turning of the seasons and the treasures each season brings. Encourage the children to bring seasonal treasures from nature that they find in their yards or on walks or hikes or even on the way to school. If you live in a more urban area, be sure to take them to places where they have a chance to gather seed pods, nuts, colored leaves, special stones, wildflowers and any other gifts from the Earth that they can find. Going back to the same places at different times of the year allows them to really experience the changes the seasons bring.

Other things that can be part of your Season's Garden are seasonal produce—also a gift from the Earth—and things that you make from natural materials.

Consider including a small garden within your Season's Garden. A plant saucer, suited to the size of your garden, with some kind of protective surface between it and your cloths or table to protect from dampness, works very well as a place for growing things. Place some small rocks or pebbles in the bottom of the saucer for draining and fill it with soil. Then it's ready to accept whatever you'd like to plant, from bulbs (fall) to moss (winter) to seeds (spring). You can also add a nice rock or two, a small piece of bark or an interesting branch or piece of driftwood. Individual seasonal suggestions accompany each Season's Garden section of this book.

Caring for the Season's Garden means keeping it neat and beautiful. It should not become a junk corner. This will mean sorting through it occasionally, removing some objects when there are too many and organizing the remaining objects so that they are easy to see and appreciate. Water the plants and seeds, put faded flowers in the compost and, generally, just tidy up. Watch for things like milkweed pods which can suddenly surprise you by opening up and sending their silky contents floating around the room. As always, gently remove any little insects to the outdoors.

Natural objects that you remove from the Season's Garden can be used elsewhere. Rocks can go into a basket to be used for building (the Whole Earth Home or Classroom section in the winter chapter describes ways to use natural materials indoors); seed pods, etc., can be saved for projects. Especially nice or unusual items can be stored for next year's garden.

Sharing the Season's Garden

Encourage the children to come and play by or in the Season's Garden, exploring its treasures, but a good guideline to follow is that things stay in the garden and don't travel to other areas of the room. The Season's Garden is a special place, and we want to keep it that way.

You can have a special time each day—during circle time might be good—when things that have been brought for the garden can be placed there, perhaps with a brief story about how or where it was found. The idea is not to study things scientifically at this point but to enjoy and, especially, appreciate them.

The Season's Garden in Fall

Colors: Muted reds, red-browns, yellows, oranges and earthy greens.

Objects: Seed pods of all kinds, acorns, nuts, Indian corn, fall flowers, weeds, pressed leaves and fall produce (pumpkins, gourds, squash, apples, pears).

Special Additions: Make a Harvest Wreath with the children to hang over or place near your Season's Garden. Start with a straw wreath form (available at craft shops) and attach sheaves of wheat and pressed leaves to it. Hang apple slices for drying (see the Apple Drying activity in the this chapter) and small bunches of washed seedless grapes from the wreath. If you dip the bunches of grapes in boiling water it will hasten the drying process by breaking the skins and sterilizing them against mold. Hang the wreath up with red ribbon, and add to it as autumn goes along. Don't forget to eat your dried apples and grapes (raisins) for snack one day.

This wreath can become a permanent part of your Season's Garden, changing with the seasons. It also represents the circle of the year.

Plantings: If you have an actual indoor garden, plant small bulbs like crocus in it. You could have a little fall festival and let the children do this ceremoniously—tucking them in the Earth to sleep through the winter.

Leaf Banners

Capture a bit of fall and make a beautiful banner that the children can display.

You will need

- work table covered with an oil cloth (available at fabric, hardware and variety stores)*
- natural colored cotton muslin cut into 8 inch squares (use pinking shears to prevent unraveling)
- several large well-formed leaves
- pen
- old toothbrushes
- jar lids
- red, yellow and brown non-toxic tempera paints**
- small branches (1/2 inch in diameter) cut into 6-8 inch lengths
- tacks or stapler
- yarn, string or embroidery floss

What to do

1. Let two or three children work on their banners at one time.

2. Write the child's name in the bottom corner on the back of the cloth.

3. Have the children place a leaf on the muslin and splatter paint around the edge of the leaf by dipping a toothbrush in paint and scraping it over the edge of a jar lid.

4. Being careful not to move the leaf, splatter paint all the way around the leaf, lift it up and see the imprint left behind.

5. Attach the banner to the branch with tacks or staples. Tie a piece of string to each end of the branch.

6. Allow to dry and use as is, or repeat the above steps the next day on the other side to make a double-sided banner.

* Oil cloth (often made of vinyl today) is a protective table covering that can be wiped clean and reused over and over. You could use newspapers, but they aren't aesthetically pleasing as a work surface, the ink comes off on your hands and on the cloth, and you have to throw them out anyway. Better to recycle the papers and use a reusable!

**Non-toxic tempera paints are those that are water-based and, generally, felt to be safe for use with children. They are available in art and school supply stores. Look for the AP (Approved Product) and CP (Certified Product) non-toxic seals on the labels. These are products which have been evaluated by the Arts and Crafts Materials Institute. They will send you a list of non-toxic art supplies. Send a self-addressed stamped envelope to:

> The Arts and Crafts Materials
> Institute
> 715 Boyleston Street
> Boston, MA 02116
> 1-617-266-6800

However, just because they are non-toxic doesn't mean they are earth-friendly. Use them as little as possible, and focus on using watercolors from plant sources for other painting experiences. See the spring Whole Earth Home and Classroom section for a description of watercolor painting with children.

Wheat Weaving

Braided wheat straw decorations—often called "corn dollies"—are symbols of good luck and prosperity. They are part of the harvest celebrations of many lands.

You will need

- wheat on the stalk—from craft supply stores or from farmers in your area who grow wheat

- tub of water

- red yarn (wool or cotton) or embroidery floss

What to do

1. Names of farmers in your area who grow wheat are often available from the local county extension service or 4-H clubs. You could also grow the wheat yourself in a sunny garden space! It's not difficult to grow.

2. Soak the wheat stems (be sure the seed heads are out of the water) in the tub of water for about an hour before you want to use them.

3. The children work in pairs with one holding the seed heads and the other one braiding. Give one child in each pair three stalks of wheat. Help the children start braiding the stems, beginning at the end near the seed heads.

4. Do a simple braid. Chanting the following helps.

Put the right one in the middle.
Put the left one in the middle.

Repeat the chant as needed.

5. Make the braid fairly tight, and braid the entire stem.

6. Curve the braided end around, overlapping it with the seed heads and tie with a yarn bow.

7. These make lovely additions to the Season's Garden or decorations for the snack table, especially at a harvest celebration. They also make great napkin holders!

8. Play a weaving ring game with the children. Stand in a circle, holding hands, and chant or sing (just make up a simple tune) the following:

Harvest crown, harvest crown
Now we weave a harvest crown.
(Susie) weaves and (Nathan) weaves
Now we weave a harvest crown.

Name two children standing side by side. When the children are named, they cross their arms in front of them and re-make the circle, holding hands with crossed arms. Gradually sing your way around the circle, calling pairs of children to weave until you've "woven" the whole circle (all the children have crossed their arms and are holding the hands of the children next to them.) After the last pair sing or chant:

Harvest crown, harvest crown
Now we've woven our harvest crown.

This a very simple, little game, but the children never cease to wonder at the weaving of it. The little ones may need help crossing their arms, and be sure to stand close enough together in the circle so that the children can reach each other.

Leaf Crowns

This is a nice activity to do while you are outdoors with the children.
Sit and make the crowns while they play, offering them to those who ask.

You will need

- lots of recently fallen leaves with stems

- a basket to hold the leaves

What to do

1. Spend some time gathering the leaves you will use. The children will want to help you with the gathering.

NOTE: Very dry, brittle leaves will disintegrate very quickly. Use fairly large, freshly fallen leaves.

2. Sit in a place where you can see the children playing and where they can see what you are doing.

3. Take two leaves and remove the stem of one at its base. Overlap the tip of one leaf and the base of the other and use the stem to attach the two leaves by pushing it down through the place where the two leaves overlap and back up again, like a straight pin.

4. Continue to attach the leaves to each other in this way. Try the crown on a child's head and when it is big enough, attach the last leaf to the first in the same way. A crown!

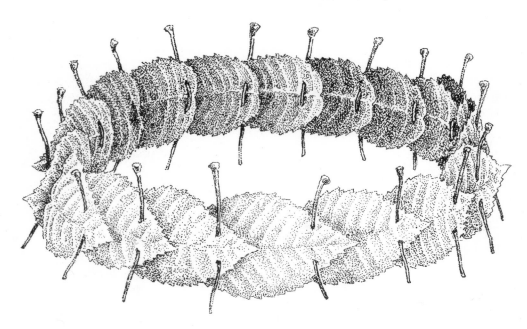

5. Continue to make the crowns for other children. Encourage the children's imaginative play by addressing them "in character": "Leaf Family," "Prince Fall," "Queen Autumn," etc.

6. Make leaf crowns or garlands to decorate your classroom.

7. The children join in the activity by bringing you more leaves, removing the stems or, with the older children, making their own crowns with assistance, if needed.

Nature's People

These little characters take on their own personalities depending on the materials used and the clothes you add. They are also very handy for "peopling" your Season's Garden.

You will need

- various kinds of nuts in the shell—filberts (also called hazelnuts), acorns, walnuts, etc.

- pine cones, nut casings, acorn caps, etc.—whatever is available in your area

- tacky glue is good because it is fast drying—or use white glue or carpenter's glue

- bits of wool felt for clothing*

- scissors

- black felt pen

What to do

1. Select two nuts, pine cones, etc., and glue one on top of the other to make a head and body. Position them according to the "character" you wish to create. For example, filberts have pointy ends that make a nice face. Some nuts have flat tops that are good places to attach heads. Acorn "tops" make good caps. Some pine cones have flat bottoms; these make good bodies.

2. Add a scarf, hat, shirt, cloak, etc., by gluing on bits of felt.

3. Use a felt pen to add dots for eyes, but do not draw in all the features. Leave that to the children's imagination.

4. Play with these little people in the sand box or in little towns or scenes the children create. They add a lovely touch to your Season's Garden, and they are delightful peaking out of a table's centerpiece.

*Use felt which has as high a wool content as you can find since it comes from nature and is much sturdier. It is also nicer to work with since it is not as stiff as the acrylic variety. Wool or wool/rayon felt is available at fabric stores and mail order from Central Shippee, Inc. and Mountain Sunrise, manufacturers of felt in 100% wool and wool blends. (See the Resource section at the end of this book for addresses.)

Lanterns

As the days grow shorter, we all need to ready ourselves for the cold days and long nights of winter. The lantern is a symbol of this: each person carrying his or her own light. Send the lantern home to be used as a lovely autumn centerpiece or the focus of a special nighttime lantern walk.

You will need

NOTE: As with all activities involving candles, children should never use their lanterns unsupervised. Also, never leave lanterns burning when you are not nearby.

- heavy construction paper or heavy white watercolor paper approximately 9 x 12 inches or 12 x 16 inches. Size is not crucial, so you can experiment a bit.

- crayons or liquid watercolors and paintbrushes

- scissors

- stapler

- tacky glue, white glue or glue sticks

- paper fasteners or 1/4 inch wire (florists' wire)

- warming candles, votive candles or tea lights (available at grocery, hardware or kitchen stores)

- masking tape

- optional—tissue paper

What to do

1. Have the children decorate their lantern papers by coloring or painting them. Suns, moons and stars are appropriate decorations, but the younger children will just make color designs. Let the children choose how to color or paint their lanterns. Encourage them to fill the paper with color.

2. Make a fold all the way across the length of the paper, approximately three inches up from the bottom.

3. Cut a fringe of three-inch wide segments all along the bottom folded section.

4. Cut several small shapes out of the top portion of the lantern. These are the "windows" that the light shines through; they can be circles, random shapes or shapes cut in the sun, stars and moon motif. The sun can be just a circular cutout with snips or colored "rays." The older children (4+) can cut their own shapes,

though they may need help starting the cuts. Small pieces of colored tissue paper glued over the inside of windows create a beautiful effect.

5. Form the lantern paper into a cylinder, stapling it at the top and bottom. Fold the fringed edges in and overlap them to make the lantern's bottom. Put small dabs of glue between the fringes to hold them together.

6. Add a fairly long handle (12-15 inches) of either 1/4 inch wire poked through the sides and twisted back up onto itself to secure it or cut 1/2-inch wide construction or watercolor paper strips and attach them to each side of the lantern with paper fasteners.

7. For the light use votive candles, warming candles or tea lights that come in individual metal cups. A loop of masking tape placed in the bottom of the lantern holds the candle in place.

Lanterns Too

This is a lovely pressed leaf lantern that is more transparent and a bit more fragile than the other style.

You will need

- pressed leaves of various kinds

- wax paper

- scissors

- round box tops—oatmeal containers and round cheese boxes work well (ask parents to send them in for you)

- iron and ironing board

- construction paper strips one inch wide in fall colors

- white glue

- florists' wire (1/4-inch wire) or 1/2-inch wide paper strips and paper fasteners

- warming candles, votive candles or tea lights

- masking tape

What to do

1. Begin collecting the leaves two to three weeks ahead of time, and send a note home asking parents to save round box lids for you.

2. Press the leaves by placing them in the pages of a heavy book (phone books work well) or in a flower press. The children love to help with this. They will be pressed in two to three days. Press enough leaves so that there are several for each child.

3. Each child selects a box lid and cuts two sheets of wax paper to fit around the outside of the lid. The lids are the bottoms of the lanterns, the wax paper the cylindrical sides.

4. Each child chooses pressed leaves and arranges them on one sheet of wax paper.

5. Cover the leaves with the other piece of wax paper the same size and, using a warm (not hot!) iron, gently but firmly press the wax paper, melting the wax and ironing the leaves between the sheets of paper.

6. Glue a one inch wide strip of construction paper (fall colors: orange, red, gold, yellow, etc.) along the top of the wax paper. This reinforces the top when you attach the handle and gives the lantern a more finished look.

7. Glue the bottom edge of the wax paper to the outside of the box lid. The easiest way to do this is to run a bead of glue all along the outside of the box lid. Then roll the lid along the bottom edge of the wax paper, forming the cylinder as you roll. You may also add a finishing strip of construction paper along this bottom edge.

8. Put several dabs of glue along the overlapping edges of the wax paper sides to close the cylinder.

9. Attach a 12-15 inch handle using 1/4 inch wire poked through the sides and twisted back up and around itself several times to secure it or 1/2 inch wide strips of construction paper attached with paper fasteners.

10. Place a loop of masking tape in the bottom of the lantern and add a votive candle, food warming candle or tea light.

Milkweed Pods

Watch for these bumpy pods in the fall. Milkweed grows wild; it is often found along the road side. Collect stalks with lots of closed pods, being careful not to deplete the supply in any one area. Put them in a vase in your Season's Garden to wait for a windy day. Just make sure you watch them so they don't burst open indoors and fill your room with hundreds of silky milkweed fairies!

You will need

- milkweed pods, one for each child

What to do

1. Give each child a milkweed pod to take outdoors on a windy day.

2. Help them carefully open the pods, and encourage the children to hold them high above their heads or run with them so the milkweed fairies can fly free!

3. Circle time is a good time to enjoy the following fingerplay:

   In a milkweed cradle all close and warm,
   (Place cupped hands together)

   Little seeds are hiding safe from harm.
   (Keep hands closed)

   Open wide the cradle now, hold it high.
   (Open cupped hands, raise them above your head)

   Come along wind, help them fly.
   (Sway open hands in the air)

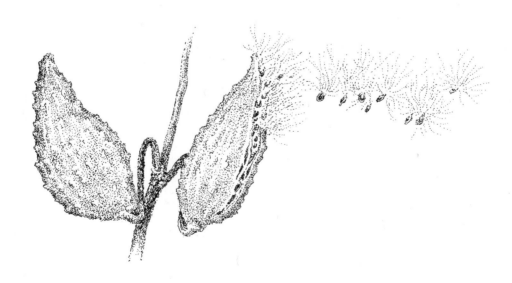

Drying Apples

Drying apples is a way to harvest nature's bounty in the fall and save it up for winter. If you have access to an apple tree or trees, all the better. (Ask parents, check in the yellow pages or with the local extension office for field trip possibilities.) Then the children can do the picking as well.

What you need

- 10-12 whole apples, organically grown if possible
- aprons for teachers and children
- sink or small dishpan
- towels
- basket or bowl
- vegetable peelers
- sharp knife for the teacher
- small knives for children—regular table knives will work but look for smaller size real knives, sometimes called luncheon knives, or buy paring knives and "pre-dull" them
- small cutting boards
- small bowl
- large needles
- thread
- optional—dowels

What to do

1. Fill the sink or a small dishpan with water and put the apples in it.

2. Let the children rinse them, dry them and place them in a basket.

3. Bring the apples to the table and peel them with the children, using the vegetable peelers (the apples dry more effectively with the skins removed).

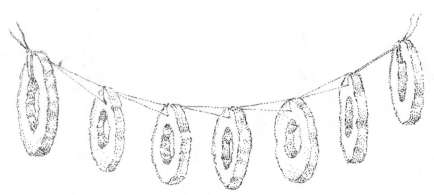

4. With the sharp knife, the teacher slices the peeled apples horizontally into 1/4 inch thick slices. Surprise! Notice the star in the center of the apple. Have the children cut around the center to remove the star and seeds. You will end up with apple slices with a round hole in the center.

Seed Saving: For all the apple activities, save the seeds that you remove from the apples, string them and dry them. They are actually quite beautiful and make lovely chains and decorations for the harvest table or your autumn Season's Garden. And, of course, as with all seeds (and all children!), they hold the wonder of what they will become. Be aware that apple seeds can be poisonous if ingested in **large** amounts. While you are saving them, keep them out of the reach of younger children.

5. Using pieces of thread about as long as your arm, thread large needles with doubled thread, knotting the thread about three inches from the end.

6. "Sew" through the first apple slice by going through the center hole and then back through the doubled thread. This will secure the apple by knotting it to the end of the thread.

7. After securing the first apple slice, have the children continue the "sewing." They sew right through the apple pulp, slide the slice down toward the last apple slice and then sew back up through the doubled thread. Leave a bit of space between each

slice so that air can circulate between them.

8. Each thread is finished when no more apple slices will fit. Knot the last slice in place as you did the first.

9. Hang the strings of apples in the room to dry. Possible places are from plant hangers; from the harvest wreath if you have one; from hooks attached to the ceiling; from a closet pole or a broomstick attached to the ceiling at each end so that it hangs horizontally. This will accommodate lots of strings of drying apples.

NOTE: It is important that the apple slices do not hang in strong, direct sunlight, so that the slices can air dry slowly. Also, each slice should be separated from the next so that air can circulate around it. If you have lots of slices that are close together, they can get moldy before they are able to dry out. If the slices are slipping together on the string, you may need to adjust them or knot them in place.

Alternative Method: If you'd like to avoid the stringing process, you can hang the apples horizontally by threading them onto a thin dowel. Hang the dowel by supporting each end. You can fit lots of apple slices on a single dowel, and even the youngest children can help.

Fancy Cutting

This is a fun thing for the teacher to do at snack time or if someone brings an apple for lunch. The children will watch in amazement and be delighted by the results.

You will need

- apples
- sharp knife

What to do

1. *Crowns:* Make sure the apple is washed. Using a sharp knife, cut apple crowns by cutting pattern one all the way around the middle of the apple. Be sure to cut well into the center of the apple but not all the way through. Hold the top and bottom of the apple. Give a twist and pull the halves apart! Surprise! Two crowns. The children will have fun putting them together and taking them apart again.

2. *Puzzles:* Cut puzzles by cutting pattern two all the way around the apple. Continue as with crowns.

3. *Stars:* Children (and many adults!) are amazed to find that there is a star inside every apple. Just cut the apple horizontally, and you will find it (number three). Enjoy this wonder by cutting the snack apples right at the table with the children, rather than cutting them ahead of time. Another way to find the three-dimensional star inside each apple (and pear) is to eat the apple whole. When you begin to get close to the core, nibble gingerly so that you get as much pulp as possible off the little core, or seed house, without actually disturbing it. Let this seed house dry and you have a three-dimensional five-pointed star. This, again, provides pure wonder for the children if you bring it to them in a "wonder full" way. It's also a great way to encourage them to eat their whole apple!

4. *Mushrooms:* Make a horizontal cut all the way around the apple, just below the center. Then make four straight cuts from the bottom of the apple down to the horizontal cut, as if you were cutting a square around the core. Take off these straight cut pieces, turn the apple right side up and you have a mushroom (number four).

1

2

3

4

Saucing Apples

Another way to preserve nature's bounty in the fall is to make applesauce for snacks.

You will need

- apples
- sharp knife for the teacher
- table or pre-dulled knives (short-handled ones if possible) for the children
- cutting boards
- large pot
- stove, hot plate or electric frying pan
- food mill
- large bowl
- spoon
- cinnamon or nutmeg
- optional—honey, brown sugar, maple syrup or sucanat (organically grown, granulated dried sugar cane)
- spoons and bowls

What to do

1. Wash and dry the apples. If they are not organically grown, you may want to peel them to remove as much of the pesticide residue as possible.

2. Using a sharp knife, cut the apples in quarters, slice out the cores and put them in the compost bucket.

3. Give the quartered, cored pieces to the children to chop and slice on their own cutting boards. The nice thing about this step is that it doesn't really matter how they cut them. Butter knives work fine for this.

NOTE: Always be sure the children thoroughly wash their hands before handling food.

4. Place the pot in the middle of the table, and put all the pieces into the pot.

5. After you have chopped all the apples, put the pot on the stove to cook on medium to low heat and add about 1/4 inch of water. If the stove or cooking apparatus is in the classroom or close by, cook the apples while the class continues, and let the wonderful aroma fill the air. Just stir it occasionally to make sure it doesn't burn. If your access to a stove is limited, other cooking alternatives include cooking the apples in a crock pot, taking the apples home to cook or cooking them at another time.

NOTE: If you use a hot plate in the classroom, take the necessary safety precautions and be very clear with the children that they may not get too close.

6. When the apples have cooked down and are quite mushy, let them cool and then bring the pot to the table with a large bowl and the food mill. Attach the mill to the edge of the bowl. Put several large spoonfuls of apples into the mill, and have the children take turns milling the apples. Every so often you will need to stop and clean the skins out of the bottom of the mill. These are great for compost if the children don't nibble them all up!

7. After you have "sauced" all the apples, season with a bit of cinnamon and/or let the children grate a bit of nutmeg into the sauce. Occasionally, if you used only tart apples, you may want to add a little sweetener (honey, brown sugar, maple syrup, sucanat).

8. Enjoy the applesauce for snack. It is an especially nice treat when made for and served during your harvest celebration.

Baking Whole Apples

Baking apples whole or turning them into cake are two other delicious ways to bring autumn's harvest into the classroom.

You will need

- large apples (preferably organic so the skins can be eaten)
- goodies for stuffing: chopped walnuts, raisins, date pieces, currants, etc.
- cinnamon or nutmeg
- sweeteners: honey, brown sugar, maple syrup or sucanat
- large bowl
- mixing spoon
- apple corers
- large baking pan (Pyrex works well)
- oven
- optional-milk or cream
- spoons and bowls

What to do

1. Wash and dry the apples.

2. Mix the stuffing ingredients with a bit of cinnamon and/or fresh grated nutmeg and a small amount of sweetener.

3. Have the children use the corers to core the apples.

4. After putting the cores in the compost bucket, stuff each apple with the stuffing mixture. Children love to do this, and even the youngest can help.

5. Place the apples in a baking dish and add about 1/4 inch of water.

6. Bake the apples in an oven at 350 degrees for about an hour.

7. Serve in bowls with a bit of milk or cream if you like. A hearty and delicious snack!

Baking Apple Cake

This tasty cake is easy to make and uses the bounty of autumn's harvest.

You will need

- 1 cup melted butter or margarine, or oil, plus extra for greasing the pan
- 3/4 cup maple syrup or honey (or some of each)
- 1 teaspoon vanilla
- 2 cups whole wheat flour
- 1 1/2 teaspoons baking powder
- 1 teaspoon salt
- 1 1/2 teaspoons cinnamon
- 1/2-3/4 cup raisins
- 1 cup chopped nuts
- 4 medium apples, peeled (if not organic), cored and chopped small
- mixing bowls and spoons
- measuring utensils
- 9 x 13-inch baking pan
- oven
- optional—paper doily, powdered sugar, sifter

What to do

1. Preheat the oven to 350 degrees.

2. Lightly grease the baking pan.

3. Mix the melted butter, sweetener and vanilla.

4. In another bowl, mix the dry ingredients.

5. Combine the liquid and dry ingredients, stirring until blended. Do not over mix; it makes the cake dry.

6. Stir in raisins, nuts and apples.

7. Spread the batter in the baking pan.

8. Bake at 350 degrees for 45 minutes or until done.

9. For an extra special occasion, once the cake is cool, decorate the top by placing a paper doily on top of the cake and sifting powdered sugar over the doily. Carefully remove the doily, and you have a lovely design.

Threshing and Winnowing

Children eat bread everyday, but many have no idea where it comes from. Allowing them to experience the whole "story" of the loaf of bread puts them more closely in touch with nature and gives them a sense of the wholeness of life. The threshing and grinding are activities that can each be done for a series of days. For example, thresh each day for a week, and then grind each day the next week—or even longer! Setting up this activity during free play time allows the children to work while they play.

You will need

- wheat on the stalk—contact your local 4-H or cooperative extension service and ask for the names of farmers in your area who raise wheat*

- large bowl or basket

- small wooden bowls

- threshing tools—sections of branches about 3 inches long and 2 inches in diameter or small round cylindrical wooden blocks

- jar

What to do

NOTE: Do **not** use treated wheat for any of these activities. Wheat prepared for commercial planting is often treated with fungicide.

1. Have the children help you pick the grain heads off the wheat stalks and collect them in a large bowl or basket. Save the wheat straw (stalks) for decorating or add them to the compost pile.

2. Work with two or three children at a time. Put several heads of wheat into a small wooden bowl and give each child a little threshing tool. Have them gently pound the wheat in the bowl with the thresher until all the grains are separated from the stalk and the outer seed covering has come off. The children can winnow the wheat by blowing gently across their bowls to blow the chaff (seed husks) away. The heavier grain stays in the bottom of the bowl.

3. Pick out the grains and put them in a jar.

4. Continue threshing over a number of days. The children enjoy doing this rhythmic activity again and again. Take turns throughout the week so that all the children have a turn.

Alternative Method: An alternative method of threshing wheat that is more like the original way grain was separated from the stalks—using long flails on the threshing floor—is to place the seed heads on a large clean sheet spread on the floor and give four or five children 1-2-inch diameter sticks about 2 feet long. The children beat the grain rhythmically to separate it from the stalks and to remove its outer covering. Be sure the beating and sticks are carefully controlled. Then another group of children, the "gleaners," pick up the wheat grains before threshing begins again. To winnow large quantities of wheat, pour the contents of the threshing bowls or sheet into a large, flat basket (a large round wooden tray with a one-inch lip also works

well) and go outdoors. Toss the wheat a little way up into the air, and the chaff will blow away, leaving you with the heavier grains of wheat. A light breeze makes it work especially well. It's really fun once you get the hang of it, and the children love to watch the chaff fly away. This can be done at outside play time.

* In the mid-Atlantic area, wheat is harvested in the summer. You can also grow it yourself without much difficulty, if you have a sunny garden plot. Just check with your local extension service about which varieties grow well in your area and when to plant them. The stalks of wheat make a lovely autumn display for the Season's Garden. Organically grown wheat berries (kernels) are available from natural food stores or by mail from Walnut Acres. (See Resources section.)

Grinding Wheat

Now that you have the grist for the mill, let's make flour! Mills are available just for flour, but these are often expensive. A fine alternative is an old-fashioned hand coffee mill. These are available at coffee or specialty stores and are not too expensive. You can also experiment using large smooth stones—one flat and one more rounded; however, this takes a longer time and produces a coarser flour.

You will need

- a hand coffee mill available at coffee and specialty stores

- grains of wheat from the previous activity or wheat berries*

- a spoon or scoop

- wide mouth jar

What to do

1. Put the wheat into the hopper with a large spoon or scoop. Hold the mill between your knees to steady it. The children turn the handle to grind the wheat. Usually these mills have a little drawer at the bottom to catch the flour.

2. The child grinds the wheat and pours the flour wheat into a wide mouth jar. The flour will be a little coarse, but that is okay.

3. The next child fills the hopper with the wheat and begins again. Continue until all the wheat berries are ground, and you have a good supply (several cups) of flour. This activity can continue for several days, or even weeks, during free play time. You never have too much flour, and the grinder then becomes a meaningful part of your Season's Garden.

4. A little verse to be chanted while grinding:

> *Miller, miller grind our grain,*
> *Grown from sun and earth and rain.*
> *With this grain we bake our bread,*
> *With this grain our friends are fed.*

* Organically grown wheat berries (kernels) are available from natural food stores or by mail from Walnut Acres. (See the Resource section.)

Baking Bread

Now, after all the threshing, winnowing and grinding, you are ready to bake bread. Have the children do most of the following with your help. And, perhaps, as the bread bakes, tell the story of "The Little Red Hen."

You will need

- 1-2 tablespoons active dry yeast
- 1/3 cup plus 1 teaspoon honey
- warm water
- small mixing bowl
- 1 teaspoon salt
- 1/3 cup oil, plus a little extra for greasing the bowl and pan(s)
- 6 cups whole wheat flour from the grinding activity, supplemented as needed
- two large bowls
- wooden spoons
- measuring spoons and cups
- wooden board or clean table top
- clean cloth
- two loaf pans or baking trays
- oven

What to do

1. In a small bowl, mix the yeast, 1 tea-spoon honey and 1/2 cup warm water. Allow this mixture to sit until it gets bubbly—approximately ten minutes.

2. In a large bowl, mix 1 1/2 cups warm water, 1/3 cup oil, 1/3 cup honey and salt.

3. Pour the yeast mixture into the large bowl and stir in three cups of whole wheat flour. Mix well, and continue adding more flour until the dough is fairly stiff and not sticky (depending on the humidity, type of flour, etc., you may need to add up to three more cups of flour).

4. Turn the dough onto a lightly floured board or clean table top and begin to knead, firmly pressing the dough away from you with the heels of your hands, folding it back onto it-self and pressing it away from you again. Continue rhythmically knead-ing the dough until it becomes smooth and elastic. Give the children small balls of dough to knead. Just break off some dough for each child who wants to help, and roll all the pieces back into one big ball when the kneading is finished. The chil-dren often like to have a turn knead-ing the big ball of dough!

5. Place in an oiled bowl, cover with a clean cloth and let rise on a sunny window sill or in another warm place until it doubles in bulk (about 45 min-utes).

6. Punch down (press down two or three times firmly but gently with your fist). Shape into two loaves or 24 small rolls (approximately). Cover and let rise once more for about 20-30 minutes.

7. Bake at 350 degrees for 45-50 min-utes for bread or 20 minutes for rolls.

NOTE: You can leave out one or both risings (steps five and six) if you want to eat the bread or rolls for snack the same day. Just have the children roll handfuls of dough into balls and place them on an oiled cookie sheet. Bake them right away, and you'll have warm, fresh rolls in 20 minutes. Make sure to let them cool a bit before eating. The children love to break the rolls open and watch the steam come out.

Harvesting Pumpkins

Pumpkins offer the possibility of many activities besides just carving them into jack-o'-lanterns. It is environmentally sound to give the children the experience of using all the parts of the whole. This is also an excellent opportunity for a field trip, and offers those in the city or in the suburbs a chance to get out to the country for a day. Check with your local extension service or with one in a county near your city about farms where you can pick your own pumpkins. Some farmers may even offer a hayride to the pumpkin patch. Make sure to arrange for a permission slip to be signed by the parents (see the Berry Picking activity in the Summer chapter for a sample), and ask that they have their children wear sturdy shoes and clothing that day—frilly dresses aren't farm appropriate. Our only pumpkin patch rules were that the children needed to stay with their group and grown-up, and carry their pumpkin by themselves. (This was to keep them all from wanting the most gigantic ones which, of course, the teacher would need to carry!)

You will need

- a local pumpkin farm
- signed permission slips
- transportation
- enough parents to accompany you so that each adult is responsible for no more than four to six children
- picnic snack or lunch to enjoy at the farm
- markers

What to do

1. Send a note home to the parents to inform them of your plan, to get their permission for their child to go and to collect any money, if necessary.

2. Arrange to have enough adults accompany you so that each adult is responsible for no more than four to six children. If using private cars, each driver should have a map and directions, phone number of the farm and a list of the names of the children in their group.

3. When you arrive at the farm, the children pick out pumpkins, and the adults write each child's name on the bottom of his or her pumpkin.

4. Eat your snack (or lunch) and enjoy the farm!

Carving and Seed Saving

Carving jack-o'-lanterns is an American Halloween tradition. Have the children help you make the class jack-o'-lantern a few days before Halloween and use it for your celebration centerpiece.

Pumpkin, pumpkin round and fat
Turn into a jack-o'-lantern just like that!

You will need

- pumpkin
- small, sharp knife
- large spoon
- large bowl
- colander or strainer
- baking sheet
- salt
- oven
- envelope

What to do

1. Cut a ring around the top of the pumpkin and remove the top. You may need to cut through strands of pulp to get it off.

2. Have the children scoop out the pulp and seeds with a spoon or their hands. Make sure to get the inside of the pumpkin well cleaned. Save the pulp and seeds in a bowl.

NOTE: Always require the children to wash their hands thoroughly before handling any food.

3. Carve a face with two eyes, a nose and a mouth. It's best for young children to have a friendly pumpkin face, so the pumpkin is not frightening.

4. Separate the seeds from the pulp. Rinsing in a colander or wire strainer helps. Place the seeds on a baking sheet, keeping a small handful to air dry for planting in the garden next spring. The stringy pulp can go into the compost bucket.

5. Sprinkle some salt on the seeds to be roasted, and bake them in a 350- degree oven for 5-10 minutes or until lightly browned.

6. The seeds you save for planting should be air dried, out of direct sunlight. After a week or so, place them in an envelope, mark with the date and contents and save it in a cool, dry place until spring.

NOTE: If local regulations allow you to have a candle in your jack-o'-lantern, use a votive candle or a warming candle that comes in a little metal cup. They don't tip over easily.

Cooking

Now let's turn our jack-o'-lantern into cooked pumpkin puree to make a delicious pumpkin pie! Do as much as possible with the children, although you may need to do the cooking down (step 6) after class or at home.

You will need

- your jack-o'-lantern and other cleaned pumpkins you want to cook
- sharp knife for the teacher
- vegetable peelers
- knives for the children
- cutting boards
- large pot
- hot plate, stove or electric frying pan
- fork
- food mill
- large bowl

What to do

1. Cut the jack-o'-lantern into large chunks.

2. Using the vegetable peelers, have the children help you scrape the pumpkin rind off the flesh.

3. Cut the peeled pumpkin into smaller chunks and place in a pot with about 1/4 inch of water.

4. Cover and cook over low heat until the pumpkin is soft. Check it by piercing with a fork. Stir occasionally, and add more water if necessary to prevent sticking.

5. Cool the pumpkin, place the food mill over a large bowl, fill it with pumpkin and turn the handle to puree the pumpkin.

6. Return the puree to the pot and cook it down over low heat, stirring often. You want the water to cook off and the puree to thicken.

7. This puree freezes well if you won't be using it soon or have extra.

Baking Pies

Now you can turn the puree into a pumpkin pie!

You will need

- an unbaked pie shell
- 1 1/2 cups pumpkin puree
- 1/2 cup honey
- 2 teaspoons grated orange rind
- 1/2 teaspoon each, cinnamon and cloves
- 1/2 teaspoon salt
- 1/2 teaspoon vanilla
- 1/4 teaspoon each, nutmeg and ginger
- 2 eggs
- 1 cup cream or half-and-half
- mixing bowl and spoon
- measuring utensils
- oven
- plates and forks

What to do

1. Prepare an unbaked pie shell with a recipe of your choice, preferably one made with whole wheat flour.

2. In a large bowl, mix pie ingredients in the order given.

3. Pour into pie shell.

4. Bake at 425 degrees for 45 minutes.

5. The recipe can be doubled to make two pies.

6. This is a very rich pie, so serve small slivers; it makes a special treat for a Thanksgiving celebration.

Removing Kernels from the Cob

Not just a beautiful autumn decoration for your classroom or harvest table, Indian corn can be the source of many harvest activities and a great example of the Native American practice of using all the parts of the whole.

You will need

- ears of Indian corn, the more the better

- large wooden bowls or baskets

What to do

1. Take one ear of corn and remove the husks. Save these in a large basket. They will be used for making dolls in another activity.

2. Take the ear of corn, hold the fat end in your hands and begin to push the kernels off with your thumbs. Work over a large wooden bowl or basket that will catch the kernels as they pop off. Continue until the cob is empty.

3. The children love to do this and usually just need you to start it for them by removing a few kernels.

4. Be sure to save all the husks and cobs for other activities.

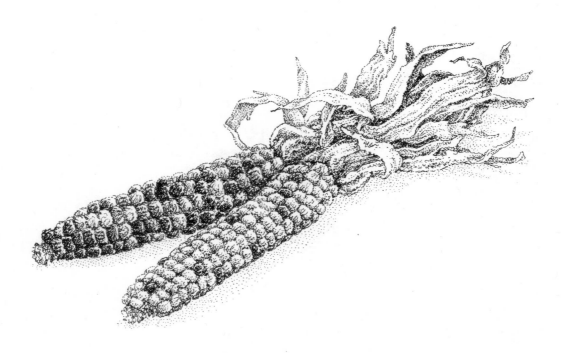

Stringing Necklaces

These easy-to-make necklaces are autumn treasures.

You will need

- Indian corn kernels
- large bowl
- water
- long strands of heavy thread (button-hole twist or embroidery floss)
- heavy sewing needles
- masking tape and pen
- towel

What to do

1. Soak a large quantity of the corn kernels in a bowl of water overnight. It's fun for the children to help you set this up the day before you plan to string the necklaces. If you plan to grind the corn (see the corn grinding activity in this chapter), save several cups of kernels for that—they will grind more easily if they haven't been soaked first.

2. In the morning, pour off the water and bring the bowl of corn to the work table. The children love to dip their hands in and "finger" the damp corn that has softened and swelled a bit overnight.

3. Give each child who is ready to work a needle threaded with 24 inches of doubled thread. Tie a knot about two inches from the end. Using doubled thread will keep the thread from slipping off the needles.

4. The children choose the kernels they want, push the needle through the center of each kernel (be careful for little fingers here) and slide the kernel down to the knot. It's helpful to start with the kernel resting on the table, rather than holding it, and just push the needle down through the kernel.

5. The children thread as many (or as few) kernels as they want, again leaving about two inches of thread at the other end. To continue this activity over a few days, just mark each string with the child's name(masking tape works well) and keep the unused corn kernels moist by covering them with a damp towel and refrigerating them so they don't get moldy.

6. When the necklaces are finished, center the corn kernels if they aren't a full string and tie the open ends together in a bow around the children's necks.

7. Some of the corn may begin to sprout—a perfect opportunity to grow some corn plants indoors to be transplanted to your garden in the spring. Then you can harvest your own Indian corn next fall.

Grinding Corn

Grind the corn with a hand coffee mill, just as you did the wheat (see "From Wheat to Bread: Grinding" in this chapter). If you have soaked the kernels already for the Indian Corn Necklaces, let them dry out for a few days before you try to grind them. If they are too moist, they will gum up the works!

You will need

- several cups of kernels of Indian corn

- spoon or scoop

- hand mill—a coffee grinder works well, available at coffee, specialty and kitchen stores

- a jar to store the corn meal

What to do

1. Put the corn kernels into the hopper of the mill with a large spoon or scoop. Hold the mill between your knees to steady it. Have the children turn the handle to grind the corn. Usually these mills have a little drawer at the bottom to catch the ground corn.

2. Let the children grind the corn and pour the ground corn into a large jar. The corn meal will be a little coarse, but that is okay.

3. The next child fills the hopper with corn and begins again. Continue until you have a good supply of corn meal.

4. Compare the ground corn with the ground wheat. They have different textures, colors, smells.

5. Try grinding the corn with large, smooth stones—one flat, one slightly rounded.

6. When you have a good supply of corn meal, bake some corn muffins.

Baking

Corn bread or corn muffins make a wonderful treat for a harvest or Thanksgiving celebration, especially when the children have done the grinding, mixing and the baking!

You will need

- 1 cup yellow corn meal from the grinding activity, supplemented as needed

- 1 cup unbleached white flour

- 1 teaspoon salt

- 3 teaspoons baking powder

- sifter

- measuring utensils

- mixing bowl and spoon

- 2 teaspoons sweetener—honey, maple syrup, sucanat

- 1 cup milk

- 1 egg, beaten

- 1/4 (1/2 stick) cup unsalted butter or margarine, melted, plus extra for greasing the muffin pans

- muffin pans

- oven

What to do

1. Grease the muffin pans.

2. Sift the dry ingredients into the mixing bowl, and then add the sweetener.

3. Add the milk and beaten egg and mix until smooth.

4. Mix in the melted butter.

5. Pour batter into the muffin pan, filling each cup 1/2-3/4 full.

6. Bake at 375 degrees for 12-15 minutes, or until lightly browned.

7. Makes 12 muffins.

Making Corn Husk Dolls

It's nice to use up those leftover corn husks by making a doll or doll family (depending on how many husks you have) for the children to play with in class or to grace your Season's Garden. The children will enjoy helping you make these.

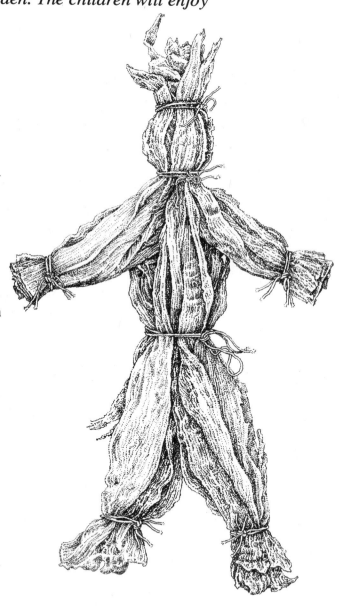

You will need

- corn husks (saved when you re-moved the kernels from the cobs)

- pan of warm water

- heavy thread—buttonhole thread or embroidery floss works well

- scissors

- optional—stuffing wool or cotton and material scraps for clothes

NOTE: If you would like each child to make a doll, you'll need lots of corn husks, and the children will need help tying the knots.

What to do

1. Soak the corn husks in warm water for about 1/2 hour.

2. Tie twelve husks together tightly at the top.

3. To make the head, tie a neck a short way down from the top. You can stuff a little wool or cotton in here if you like.

4. Separate three husks on both sides, and tie them halfway down for the arms. Trim the excess.

5. To make the body, tie the remaining husks just above half way down.

6. Make legs by tying three husks on each side, a bit up from the ends. Trim the ends.

7. Add clothing—scarves, hats, shawls, skirts, jackets, etc.

8. Make corn husk children by starting with shorter husks.

Grating the Cobs

This is an activity that children enjoy doing, and often occupies them during free play.

You will need

- Indian corn cobs

What to do

1. Place the Indian corn cobs in the playhouse area.

2. When two cobs are "grated" together, that is, one is rubbed against the other, a fine, flaky powder is produced that children love to save in bowls for "cooking."

3. This is a very simple activity which children love to do. The rhythmic grating is fun and can occupy children for long periods of time.

4. Children will probably find lots of other uses for the corn cobs, and the cobs will become a regular part of your playhouse equipment.

Chapter 2
Winter

The Whole Earth Home and Classroom

Toys from Nature: The Indoor Play Space

Preschool-aged children are blessed with very active imaginations. For them, play is work, and through creative play, they learn about the world and how to live in the world. We can support this process of exploring and growing by providing children with toys and materials that can grow with them. Nature offers many of these materials—free for the collecting.

Many of the "educational" toys that fill our homes and classrooms are not only costly but thwart the innate creativity of the children. They are designed to accomplish a process carefully thought out by adults and, therefore, allow for little creative input by children. The best toys are the simplest—those that allow the "player" to use his or her individual imagination and those that can be used for many different purposes. For example, a basket of stones can be used for building a wall or a road; they can be stirred into a "soup" in the playhouse; they can be counted as money for the shopkeeper; they can be tokens for a ride on an imaginary train or bus; they can, in short, be used for whatever the play situation requires.

Besides the creative possibilities and the economic advantages they offer, toys from nature have the added benefit of being "real," that is, they come from the Earth and not from a factory. They are durable, beautiful and interesting in their shape, form and color. They are pleasing to see and to handle. This offers quite a contrast to Day-Glo, non-bio-degradable plastic toys. The "un-earthly" look of many commercial children's "toys" is also an environmental statement.

Building Blocks and More

Gradually begin to use more natural materials in your home and classroom. With the children's help, gather stones, shells, pine cones, anything that nature offers. Carefully wash and dry the stones, and sort other objects into appropriately sized baskets.

Bring in a few small branches and logs. Set up a woodworking bench, and let the children help you saw them into slices. These make excellent building blocks, and offer a wide variety of building possibilities, often more challenging and interesting than building with the more uniform square and rectangular blocks. Drill small holes in the sides of some of these smaller "blocks" and insert small dowels or sticks into the holes to make fences.

Someone who knows how to use a chain saw could cut you a whole basketful of these natural blocks in no time. Make them of varying lengths and try to cut them fairly flat, although those with slanted cuts offer interesting possibilities too. Sets of natural building blocks are also available mail order from Nova, HearthSong and The Ark (addresses in the Resources section). However, it is very easy to make them on your own, and it is a wonderful experience for the children. Sand and wax or polish the blocks if you like, or just use them as is. Over time they will develop a lovely patina.

Introduce these natural building materials by telling a story with them. Once you have gathered a fair amount of different things, sit down at free play time and set the scene. Spread a large cloth—preferably earth or grass colored—or build on the carpet if you have one. Stones line the banks of a little creek (a long blue cloth). The creek flows down out of the mountain (a pile of stones) and runs past a forest (a stand of pine cones) and a farm (a house built of wood pieces, surrounded by a stick fence and, perhaps, some small wooden animals). The creek flows all the way to the beach (white or sandy colored cloth with shells) and the sea (a big blue cloth). Here come a few children (small wooden figures) travelling down the creek to the sea. Just imagine (the children will!) all the adventures they will have along the way.

Little stories like this help the children get started using these new materials. If they are not used to having open-ended materials for play or if their imaginative capacities have not been nurtured, it may take a bit longer to unlock this hidden potential. This, unfortunately, may be the case if children have been accustomed to passively sitting in front of the television. But all children are born with the capacity for creative, free play. We need only be supportive facilitators.

Stumps

Another gift from nature that is very versatile and useful in the home or classroom is tree stumps. These are a little harder to come by, but even in cities, old or sick trees are cut down or trimmed, and it just takes a little effort to contact the people responsible for this kind of work—perhaps the Department of Public Works. I have fond memories of trying to roll a tree stump three feet in diameter (heavy!!) up a ramp and into the school van. A parent had told the school about tree work being done in her neighborhood. Two of us managed, somehow, to get quite a few stumps. Preschool teachers can be very persistent! The stumps, varying in size from one to three feet in diameter and from three inches to two feet in height, became a focal point of our classroom play situations and were the impetus for much real and earnest work.

We spent lots of time removing the loose bark from some of the stumps. We used rasps, files and sandpaper to smooth the rough edges. The larger pieces of bark went into a basket. They served a variety of purposes—as roofs for little buildings, as boats with leaf sails and even as loaves of bread in the "bakery."

One day we used an auger—a large hand-operated drill—and with much effort and lots of turn taking, managed to drill a hole in the center of the largest oak stump. Into this we placed a four-foot long 1 1/2 inch dowel—the kind used for closet poles. A cup hook screwed into the top of the dowel enabled us to attach a pulley so we could raise and lower a flag. A smaller dowel lashed on at the bottom allowed us to attach a sail. Many happy days were spent aboard this "ship."

Other stumps became walls, seats on buses or trains, stools and chairs in a house—any number of useful things! What a wonderful large motor activity—moving the stumps around the classroom! The smallest stumps (three to four inches thick) were often placed on top of other stumps and used as steering wheels! And other times we tied small "oxygen tank" logs on the children's backs so they could be deep sea divers. As you can see, children are very creative and resourceful when given materials which encourage the use of their imaginative capacities.

Filling your classroom (or home!) with toys from nature offers children endless creative possibilities for play. It also surrounds them with the warmth and beauty of the natural world, nourishes their senses and fosters, in a subtle yet solid way, their appreciation of all the many gifts that the Earth offers us.

Bringing Nature In:
The Season's Garden in Winter

For more detailed information about setting up a Season's Garden, see the section on "Creating a Seasonal Garden" in the fall chapter.

Winter colors: Icy blue, white, deep blues, deep violet, dark red.

Special objects: Special rocks and stones; crystals; branches of evergreens, placed in a vase of water as you would flowers; a white candle, even if you can't light it; star-shaped seed pods; tiny brass stars or stars cut from gold paper.

Wreath: Remove the autumn decorations from your wreath. Gather evergreen branches (ask parents to bring in pruned branches) and transform the wreath one day during class. With the children's help, bind the evergreens over the straw base. Use dark colored cord or heavy string to hold the evergreens in place. Using a pretty ribbon to hang the wreath also makes it more festive. Be sure to "water" the wreath daily with a plant mister to keep it from getting too dry.

Plantings: Carefully gather moss from an outdoor area (a damp area like a creek bed or the ground near the north side of trees) and transplant it in your dish garden (a large clay plant saucer). You don't have to fill the garden, even a small mossy area is nice. Mist it occasionally. Tuck in little stones and pieces of bark or tiny pine cones.

NOTE: Be sure that when you gather things from nature, such as this moss, that you take only a little from any one area, so as not to deplete the "stock" in that area. Also, remove it carefully and refill or tamp down the area so that you don't leave scars. The Native Americans always asked permission of the Earth, tree, etc., before using it. Our rededication to this attitude of reverence towards the natural world will mean a lot to our children, and they will imitate it.

Special Activity: Force paperwhite narcissus bulbs, available at garden or hardware stores. Just two or three of these lovely blooms will perfume your whole room. They are easy to grow in a small dish of pebbles. Nestle the bottoms of the bulbs in the pebbles. Pour in enough water so that the bottoms of the bulbs are always damp, and replenish the water as necessary. I used a dish of bulbs on each of two snack tables—a lively centerpiece. Start the bulbs about two to three weeks before you would like them to bloom.

Caring for the Birds and Squirrels

Caring for the birds and squirrels is a way of bringing nature very close to your classroom—especially in more urban locations. The rhythmic responsibility of caring for and feeding the birds each day is very rewarding for the children and develops good habits.

You will need

- bird seed—often available in bulk from hardware and garden shops

- bird feeder(s)—this can range from something you make or buy to a flat tree stump or window sill

- bird bath—a large shallow pan, dish or large plant saucer works as well as the standard bird bath

- nuts in shells

- dried corn on the cob—for the squirrels to eat, but they may also enjoy your birdseed

- large, thick nails

- small brush

- optional—large juice can or plastic jug and string

What to do

1. Find a place within view of your windows to set up the feeding station. It may be a window sill if you're not on the ground floor. In this case, a bird feeder that attaches to the window with suction cups may be best, or attach a small board to the window sill.

2. Birds like a protected area, so place a bush or tree (container plantings will work) in the feeding area outside your windows.

3. Hang the feeder there or place the seed on the ground or on a tree stump. Place the nuts for the squirrels on the ground or on a low stump or flat rock nearby. If you are feeding the squirrels corn on the cob, drive a large, thick nail into the stump leaving an inch or two exposed, and press the flat end of the corn cob down onto the nail head to hold it upright. The squirrels will also appreciate apple cores and carrot tops left over from snack, and they'll love your roasted pumpkin seeds left over from cooking your Halloween jack-o'-lantern. Place the water dish or bird bath nearby. It is better to place it down on the ground

a low stump than to have it up on a pedestal. This way it is more accessible to the children for filling and cleaning. Just strive to make the feeding area a pleasant place for the birds and the squirrels to be. Add plants, rocks, etc., to naturalize the setting.

4. Each morning choose a child or children to help you fill the feeder(s) with seed, put out some nuts and corn and replenish the water supply. Free play time is a good time to do this, and the children love to help. The bird bath should be emptied, scrubbed with a small brush (no soap) and rinsed once a week. This is another favorite job and a chance for some water play.

5. If the birds and the squirrels don't come right away, keep putting out the food and be patient. They will soon find it.

NOTE: One way to attract birds is with the sound of water. Punch a small hole in a large juice can or a plastic jug, fill it with water and hang it above the bird bath so that the water will drip down into it.

6. Birds and squirrels are often the wildest creatures that many children get to see other than the animals in the zoo. Caring for them regularly is a simple way of helping the children feel a connection to nature and a responsibility for its well-being, particularly in the winter months when the animals have a harder time finding their own food. If you live in an area where it gets cold, just wait for that first "wonder full" day when there is ice in the bird bath!

Pine Cone Bird Feeders

These are easy to make and the birds love them!

You will need

- pine cones—long ones work best, but any kind will do

- peanut butter—the cheapest sugarless kind you can find—many natural food stores allow you to buy it in bulk, so you can buy as much or as little as you want. Figure that you will be able to make approximately 10 pine cone bird feeders per pound of peanut butter.

- popsicle or craft sticks or tongue depressors

- bird seed—the least expensive kind you can find

- cookie sheet or a pie plate to hold the seed

- heavy string, twine or yarn

- wax paper or wax paper bags

What to do

1. Apply the peanut butter to the pine cones with the craft sticks, pressing it into all the nooks and crannies. Expect that fingers will be licked and some peanut butter will be eaten. Just remind the children that this snack is really for the birds!

2. Roll the peanut butter-covered pine cones in the pan of bird seed. The bird seed will stick to the peanut butter. Gently shake off the excess.

3. Tie the string or yarn tightly around the base or top of each pine cone.

4. Perhaps the children could make two—one for school and one for home. This way you will have a supply to use at school and the children will be able to see how they attract the birds. Then the children will bring them home enthusiastically!

5. Send them home in wax paper or a wax paper sandwich bags. They are a lot less sticky than plastic (besides being more environmentally friendly).

Star Windows:
Tissue Paper Transparencies

These little windows of colored light are a special addition to a window near your Season's Garden. They also create a seasonal mood, depending on the colors and forms you choose. To the children, they are magical when placed on the window so that the sunlight shines through.

You will need

- construction paper—deep red and blue

- scissors

- colored tissue paper—golds, yellows and white*

- basket or tray

- pen

- white glue, small saucers and cotton swabs (or pencils with new erasers) or glue sticks

What to do

1. Cut the construction paper into frames in the shape of your choice. A plain rectangle with rounded edges to soften the shape is fine; a star would be extra special, but will take longer to do. The total size is up to you, but make it no smaller than 6 inches long and 4 inches wide. The frame edge itself should be approximately 3/4 to one inch thick.

2. Run a bead of glue all around the back of the frame near the outer edge and press a large piece of white tissue paper over it. Trim the tissue paper.

3. Cut the yellow and gold tissue paper into star shapes, in varying sizes if you like. By folding the tissue accordion style, you can cut many stars at the same time. Place the stars in a basket or on a tray.

4. Write the children's names on the backs of the frames (the side with the tissue paper glued on). This is the side on which they work.

5. Give each child a glue stick or have them share a small dish or saucer of glue (just put a little on the dish). They could use a cotton swab to apply the glue, but use these sparingly as they cannot be reused. You can also use erasers on new pencils to dip in the glue and dab onto the stars. The erasers can be rinsed off and reused.

6. Using a **tiny** drop of glue for each star, the children cover the white tissue with stars. Encourage them to fill the surface, although the younger ones may just place a few stars on the tissue paper. It is fine to overlap the stars, as this creates interesting shapes and beautiful and surprising new colors.

7. When they have finished, add an extra covering of white tissue as you did in step two. While not completely necessary, it provides a nice protective backing and gives the transparency a more finished look.

8. Let the children hold them up to the light. The children are always captivated by the results.

9. Cover your classroom windows with star windows, or send each one home so the stars shine there. This technique for making tissue paper transparencies is endlessly variable. The frames can be cut into any shape, from pumpkins (fall) to eggs (spring). Just choose the tissue paper in seasonal colors and cut the construction paper into appropriate shapes. Generally, white tissue paper is used for backing in all cases.

ALTERNATIVE METHOD: You can also cut the frame on a fold. This will give you a hinged frame. Glue the white tissue to the inside top and have the children glue on their stars. Then glue the sides of the frame together and you have a very nicely finished star window.

* Tissue paper is widely available in arts and crafts and school supply stores, and is also available mail order from Chaselle, Inc. (address in Resources section).

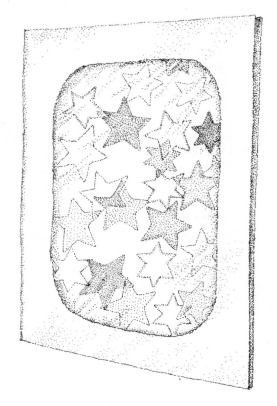

Snow Scenes

Here's a chance to play with "snow" indoors without the melting mess and create something lasting and beautiful.

You will need

- Ivory Snow soap powder

- mixing bowl

- rotary egg beater (hand-operated type)

- pieces of fairly sturdy, smooth cardboard cut into rounded, irregularly shaped pieces (approximately 4 x 2 1/2 inches)—one piece per child

- pen

- spoon

- natural objects—small stones or crystals, bits of bark or tiny sticks, tiny hemlock cones, acorns or acorn caps, holly berries or cranberries, tiny bits of evergreen, etc. Sort these into bowls or baskets.

- optional—beeswax*

What to do

1. Place about one cup of Ivory Snow in the mixing bowl and add about 3/4 cup water. Beat until the mixture holds stiff peaks; do not over beat. It should be the consistency of whipped cream. This amount should make enough "snow" for 10-12 pieces of cardboard, depending on size. Let the children help with the measuring, pouring and beating.

2. Write each child's name on the underside of his or her cardboard, and have the children choose a few objects with which to create their snow scenes. Mound one to two heaping tablespoonfuls of "snow" onto the cardboard base and spread it out a bit, leaving a little cardboard border around the edge.

3. Press the sticks, stones, etc., gently down into the snow, one at a time. Guide the children a bit as to the placement so that they are creating a woodland scene and not just a jumble. For example, a stick becomes a log on which a little red beeswax bird could perch. Then group a bit of evergreen, a stone and a pine cone or acorn around it. The idea is not to see how many things can fit on the cardboard, but to choose a few objects and place them carefully.

4. Let the snow scenes sit undisturbed for one hour, and then the children can take them home. If you have whipped in too much air, they may need to dry overnight (they will have the consistency of marshmallows).

5. Cut a larger cardboard and make a class scene to place in your Season's Garden. A small mirror placed down in the snow becomes an icy pond, a larger mound of snow a hill for sledding. Create skaters, children sledding, snowball throwers and snowmen with colored beeswax. You might even fashion a snowman from the snow. The older children could work on this for days.

* For information on molding and shaping beeswax and where you can find it, see the instructions in the activity, "Beeswax: Modeling," (page 102) in this chapter.

Nutmeg Grating

Children love to grate things. Start with whole nutmeg which will make the classroom smell like a wonderful spice shop. They will then want to try to grate other things, so let them. Save avocado pits for them to use, corn cobs and bits of tree bark work well too. The spice gratings can be saved for baking or potpourri. The corn cob and avocado gratings are great for "cooking" in the playhouse and eventually end up in the compost.

You will need

- nutmeg graters—have several as part of your equipment. They are inexpensive and are available from kitchen and hardware stores.
- grating materials—whole nutmeg, corn cobs, avocado pits and whatever else you or the children can imagine grating
- bowls
- covered containers

What to do

1. Place the grating materials and little graters in the playhouse "kitchen" along with bowls to grate into.

2. You may not even need to show the children what to do—just stand back and watch.

3. Have covered containers to save the gratings from day to day. Decide if you want to let the children take some home; they will probably want to. The children in my class liked to wrap little packages of grated nutmeg in paper towels. It was like gold dust to them.

Finger Knitting

This might more appropriately be called finger crochet as the result is a lovely yarn chain that can be made into many things or used as is. Working with warm woolen yarn is a nice wintertime activity, and finger knitting is a pleasant way to pass time indoors. The rhythmic pulling of the yarn can be quite quieting.

You will need

- bulky wool yarn*

- scissors

- a basket or cloth sack for each child's work

What to do

1. Ask at your local yarn shop for 100% wool, bulky weight yarn. Don't forget to ask for a teacher's discount! Try to get real wool rather than acrylic, as the wool comes from nature and has a much nicer, warmer feel. It's much more pleasant to work with. The younger the children, the thicker the yarn should be.

You could also use cotton yarn, but it is often not as thick and tends not to slide as easily.

2. Cut long lengths of yarn (four to five arm lengths), and have the children roll them into balls. Give each child a small handwork basket or cloth sack marked with his or her name in which the projects "in process" can be kept. This keeps the materials organized and lets the children work on the projects for days at a time. Free play time is a good time to work on this project.

3. Start the finger knitting by making a slip knot. Lay one end of the yarn across your open palm and hold the end down with your thumb. Wrap the yarn once around your fingers and **cross over** the first piece. Hold it down again at the crossing point with your thumb. Now you have a loop around your fingers. Still holding on with your thumb, slide the loop off your fingers. Bring the long side of the yarn back through this loop to make a second loop, and while still holding the short end of the yarn, pull the second loop up. This will make a knot by tightening the first loop. If the second loop gets too long and unwieldy, just pull on the long "tail" of the loop to make it smaller.

4. Continue finger knitting by reaching down through the loop to pull the yarn strand up through the loop. Always hold the finger-knitted strand firmly and near the open loop.

NOTE: It's nice to have a little verse to help the children remember what to do. Here's a suggestion:

> *Reach into the lake, (The open loop)*
> *Catch a fish to bake. (Pull up the yarn)*

5. When you get close to the end of the yarn, pull the end up through the loop to knot it off.

6. If you have a particular finished length in mind, start finger knitting with approximately four to five times that length.

7. The resulting knit "ropes" can be used for free play in many different ways. We kept a basket full of them on our shelf, and they were used for horse reins (place the center of the rope at the back of the neck, bring the two ends to the front, cross the ends and bring them under the armpits to the back); hoses for fire fighters; ties for cloth capes and other clothing; for "wrapping" presents; for building houses, making gates, etc. They are very useful.

8. You can also use the finger knitting to make more conventional items. They make perfect mitten strings. Stitch one end to each one of a pair of mittens or gloves. Run the string through the child's coat sleeves—no more lost mittens! A very nice rug can be made by coiling the finger knitting and stitching it together to make a round braided-type rug. Perhaps your playhouse or doll houses could use one. Or you could be very ambitious and make one for your story corner!

* Nice thick wool yarn for finger knitting is available from Mountain Sunrise and Nova. (See the Resources section for addresses).

Sewing Gnomes

Mythology tells us that the gnomes are little beings who help take care of the Earth, working down underground during winter months. Sewing is a wonderful wintertime activity and is suited to a variety of ages. The younger children may sew a stitch or two and go off to something else. The older ones may want to stitch lots of gnomes. Perhaps they would like to make seven in different rainbow colors and a Snow White (who doesn't necessarily have to be Caucasian!) to go with them. Sewing is a wonderful quieting activity and can be helpful for children who are at loose ends or in need of calming down. Just bring them to the sewing table, sit them down and get them started.

You will need

- felt with a high content of wool in a variety of colors: red, yellow, green, blue, violet, etc.*

- scissors

- straight pins

- large eye needles

- thread

- wool or cotton batting for stuffing**

- baskets or bags

What to do

1. Place the gnome pattern on a piece of folded felt and cut it out. You can make the pattern larger or smaller to create gnomes of different sizes. The older children can cut their own if you pin the pattern to a small square of felt.

2. Thread the needles with doubled thread, about 12 inches long (that would be 24 inches to start with), and tie a knot at the end of the thread. This will keep the thread on the needles, and the teacher from going crazy rethreading needles.

3. Using a whip stitch (over the top edge of the material), sew the tow sides together from the top of the pointy hat down to the rounded face opening; knot and cut the thread.

4. Re-knot the thread, and sew the body of the gnome.

5. Stuff generously with stuffing wool or cotton so the gnome feels firm. You can pull down a bit of wool from the bottom of the "face" to make a beard.

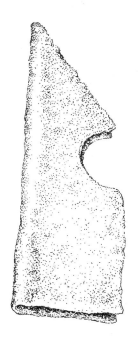 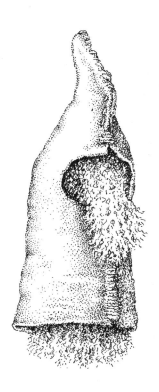

6. Don't expect the younger ones to do a perfect job. They will eventually improve their sewing skills by imitating the teacher and the older children. Let them do what they are able to do, and help them if they need help. You can fix any drastic mistakes.

7. The older children may want to make more than one gnome. Keep the gnomes in the children's handwork baskets or bags, and they can work on them for several days or weeks. This project could last throughout the winter.

* Wool felt is available at fabric stores and mail order from A Child's Dream or Nova. (See Resources section for addresses.)

**Cotton batting is often available in fabric stores. Wool batting for stuffing is available at some craft stores and mail order from West Earl Woolen Mill. (See Resources section for addresses.)

Tissue Paper Dolls

The children can complete their set of gnomes with a Tissue Paper Doll—Snow White, perhaps? This method of making the dolls can be modified for many different occasions by using different colors. The children may want to make several.

You will need

- tissue paper—white or different colors

- scissors

- stuffing wool or cotton

- string or strong thread

- optional—colored wool fleece and glue

What to do

1. Cut tissue paper into two rectangles—one 4 x 8 inches, and the other 3 x 6 inches. Use one color for both pieces or vary the colors.

2. Have the children form a bit of stuffing wool into a small tight ball about 1/2 inch in diameter. This will be for the head.

3. Place the ball just about half way down the larger tissue paper rectangle and fold the tissue over it, gathering it where the neck should be.

4. Place the smaller tissue paper rectangle over the top of the head side to side and also gather this tissue paper around the neck, leaving the front open—this will be the face.

5. Help the children tie the thread (use it doubled if it's not strong) around the neck, knotting at the back of the head.

6. Twist the two ends of the smaller rectangle into arms and hands. The back will look like a small cape.

7. Fluff out the skirt so the figure will stand.

NOTE: You may add a bit of colored fleece for hair before putting on the smaller rectangle, and you can put just a drop of glue on the places where the skirt comes together at each side to keep these from separating.

8. As noted above, this little doll can become many different characters. From the Elsa Beskow picture book, *Ollie's Ski Trip*, try making the wonderful old lady, Mrs. Thaw, who comes with her broom (a small stick to which you bind a few pieces of straw) to sweep winter away.

Yarn Dolls

These little dolls are very easy to make. The children will just need your help with tying knots.

You will need

- medium weight wool yarn in assorted colors. Thinner yarn also works, but thick yarn is too bulky for this craft project.

- scissors

- firm cardboard, corrugated works best, cut into 3 x 5 inch pieces

- optional—fabric scraps, safety pins

What to do

1. Measure out approximately 3 arm lengths (about 3 yards) of yarn and have the children roll it into a little ball. You will need one for each doll.

2. Give each child a piece of cardboard and a little ball of yarn. Have them wrap all the yarn around the cardboard. Wind the yarn around the long way for a bigger doll (almost 5 inches tall) or around the width for a smaller doll.

3. Slide a short piece of yarn under the wrappings to gather them, and tie them tightly together.

4. Now slide the wrapped yarn off the cardboard. The tie will hold the yarn together.

5. Take a small piece of yarn and tie the bundle tightly where the neck should be. This creates the head.

6. Separate about 1/4 of the strands of yarn to the left and 1/4 to the right to form the arms (just do it by sight, it needn't be exact). Tie each of these smaller bundles where the wrist should be and trim away the excess. Don't make the arms too long.

7. Tie around the body bundle at the place where the waist should be. If you want to make a doll with a skirt, you can stop here. If you want to make a figure wearing pants, separate the body bundle into two bunches, tying each off at the ankle. Trim the ends.

8. Let the children help as much as possible with the knot tying. They can also help with the trimming.

9. The older children may wish to add clothes: scarves, shawls, belts... This is a great way to use up felt and fabric scraps.

10. The 3 x 5 inch dolls make nice additions to the class toy collection and are useful when the children are building farms, villages, etc., to "people" their scenes. They may want to make several dolls. They are so simple that they leave lots of room for the imagination.

11. The smaller dolls also make sweet gifts as decorative pins. Just add a safety pin to the back, and you can pin them on hats, coats, etc.

Wooden Candleholders

These are really quite simple to make once you've gathered the materials. They are very beautiful and very practical, too. If you have made your own candles, put them in the holders and wrap them in a big sheet of tissue paper —a lovely and useful gift!

You will need

- slices of logs—one per child—approximately 3-4 inches in diameter and 1 inch thick with a 3/4 inch hole drilled to a depth of 1/2 inch in the center of each slice*

- marker

- non-toxic white or yellow glue—yellow carpenter's glue works very well

- an assortment of small things from nature: tiny pine cones (hemlock), acorns or acorn caps, tiny shells, sweet gum balls, small seed pods, nuts (especially filberts or hazelnuts) and cranberries (these provide a beautiful splash of color)—use anything that's not too large

- optional—bird seed and spoon

What to do

1. Write each child's name on the underside of the candleholder.

2. Spread the glue, fairly thickly, all over the top of the candleholder, taking care not to get it in the hole that will hold the candle. The glue will dry clear.

3. The children choose natural objects and place them all over the top of the candleholders, pressing them down into the glue. Encourage the children to cover the top. If you want, you can sprinkle a spoonful of bird seed all over the top of the candleholder when the children have finished to fill in any spaces or holes.

4. Place the candleholder in an out-of-the-way place to dry.

5. If you are making candles—either dipping, rolling or decorating—wait until those are done and send one home in its own holder.

* Perhaps you can get a carpenter-parent or local woodworking shop to do the cutting and hole drilling for you. Bring them your 3-4 inch diameter logs or branches and let them do the preparation on their equipment.

 SUPPLYING THE MISSING LINKS: GIFT MAKING

Herbal Sachets

These sachets are easy to make and are useful, beautiful gifts. Don't be too concerned about the quality of the sewing—it's thought and effort that count here. These can also be nice Valentine's Day gifts by cutting red felt into heart shapes.

You will need

- paper (for the pattern)
- scissors
- wool felt in assorted colors—available at fabric stores*
- straight pins
- sewing thread in matching or contrasting colors
- large eye needles

- sachet mix—choose a pleasant smelling one that is not too overpowering or make your own. Try herb or spice stores or bath shops.
- wool or cotton batting—mix with the sachet material to stretch it
- optional—tissue paper, ribbon, yarn

What to do

1. For each child cut out two hearts (see illustration), or whatever other shapes you want to use, for each sachet. Cutting the shape on a fold will insure that the shape will be symmetrical. Older children can help cut by pinning the folded pattern to a folded piece of felt.

2. Pin the two (front and back) pieces together with one straight pin. Double thread a needle and tie a knot in the thread ends.

3. Start the sewing by bringing the needle through the fabric from the inside so that the knot will not show.

Have the children sew around the edge using any kind of stitch.

NOTE: I found that a whip stitch which goes through the fabric and then over the top is easiest for the children to do. Its greatest dangers are stitches too far apart or pulling the stitches too tight which makes a crinkle at the edge of the fabric (this requires fixing only in extreme circumstances).

It's just important to be consistent and keep the stitches fairly small and close together so the sachet mix won't leak out. (Continued on next page.)

4. Have the children stop sewing when there are about 1 1/2 inches left to sew, so that they can stuff their sachet.

5. Stuff with the sachet mix (and wool or cotton if you are using it), and have the children finish their sewing.

NOTE: I like to combine lavender, dried rose buds and star anise (a spice) all of which are interesting, beautiful to look at and smell good! These are also easy to handle as they are (except for the lavender) large and not too flaky.

6. When they reach the end (where they began sewing), knot off the thread and check to see that there are no gaping holes around the edges that need repair.

7. The children can take these home as is or wrap them in colored tissue paper. You can also stitch a piece of thin ribbon to the top (or have the children finger knit a cord) to make the sachet into a necklace.

* Wool felt is also available mail order from Mountain Sunrise and Nova. (See the Resource section for addresses.)

 SUPPLYING THE MISSING LINKS: GIFT MAKING

Pompon Balls

Fun and easy to make, and fun to play with. The addition of a tiny brass bell or jingle bell makes them extra special gifts. This is another project that you can have the children work on over a period of days.

You will need

- wool yarn—bulky weight (thick) is best for 4 inch balls. If you use medium weight yarn, the wrapping takes longer.

- large scissors

- basket

- cardboard

- heavy thread (buttonhole or embroidery)

- yarn needle

- medium-to large-size jingle bells or little brass bells

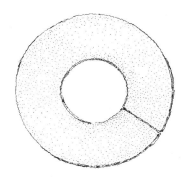

What to do

1. Prepare the yarn by cutting it into arm length strands. Keep these in a basket.

2. Cut two four inch donut-shaped forms (see the illustration) from the cardboard for each pompon. Cut a slit in each form. This will allow you to slide the pompon off when you are done, so the form can be reused by another child.

3. Give the children a set (2) of cardboard forms and an arm's length of yarn. Make the pompons just one color, or let the children choose and have confetti-colored "pot luck" pompons!

4. Have the children wrap the yarn around the form going through the center hole and around the "donut" part of the form. To begin, hold the end in place and wrap the yarn around it a few times. To finish a strand, tuck the end under the wrapped ones to keep it from unravelling.

5. Continue wrapping the strands of yarn until the center hole of the donut is completely filled up. Tuck the last end under the other strands.

6. Using a large sharp pair of scissors, cut around the outside edge of the pompons, keeping one blade between the two cardboard forms.

7. Using heavy thread or doubled yarn, tie the pompon tightly through the middle of the two pieces of cardboard.

8. Slide the yarn off the forms through the slits.

9. Fluff up and trim the pompon. Save the trimmings for future projects.

10. To add a bell, thread a yarn needle with doubled yarn and knot it. Run the needle through the bell's loop and then back through the yarn to anchor the bell to the yarn. Then thread the needle up through the center of the pompon. Pull snugly, so the bell hangs freely at the bottom edge of the pompon. Cut the thread and knot it so that it forms a large loop at the top of the pompon. Hang it on a door knocker or doorknob at holiday time to cheerfully announce the arrival of guests.

Pomander Balls

These clove-studded fruits make useful gifts and are fun to make. Hung in a closet, they delicately scent the air for up to a year. They are natural air fresheners.

You will need

- oranges, lemons or apples

- push pins or small nails

- whole cloves

- spice mixture: ground cinnamon, all-spice, cardamom, cloves—mix your own combination

- shallow pan to hold the spice mixture

- yarn or ribbon for hanging

What to do

1. If necessary (when using a thick-skinned fruit), use the nail or push pin to puncture holes in the skin of the fruit about 1/4 inch apart.

2. Press the whole cloves into the holes, or directly into the fruit, working to cover the entire skin of the fruit. This is another project which can be done over several days.

3. When the entire fruit is studded with cloves, roll the fruit in the spice mixture.

4. Knot two pieces of yarn or ribbon to-gether at the middle and place the po-mander over the knot. Take up the four ends, tie them together in a firm knot and then a pretty bow.

5. This project can be varied according to the age of the children. The five-year-olds should be able to do this by themselves. The three-year-olds may need the holes punched for them. Per-haps, rather than having each child make his or her own, the younger children can help with a "class po-mander" to hang in the bathroom or in a sunny window.

Stained Glass Triptych

This is an extra special version of the tissue paper transparencies in the winter crafts section, and uses the same techniques. These triptychs make wonderful holiday gifts.

You will need

- colored construction paper or water-color paper, card stock or poster-board for the frames

- scissors

- tissue paper—white and colors. You can use scraps.

- white glue or glue sticks

- pen

What to do

1. Make the frames by cutting the three-sided form (see illustration) from pa-per or posterboard. The approximate measurement for the center section is 6 inches high and 5 inches wide. The side measurements are 4 inches high and 3 inches wide. This will fit across an 8 1/2 x 11-inch sheet of pa-per.

2. Cut out a "window" from the center section leaving a 3/4-inch frame. You can also cut windows in the side frames.

3. Glue a piece of white tissue paper over the back of the window opening(s).

4. Give the children their frames and, working on the back side, have them **tear** small pieces, approximately the size of a quarter, of colored tissue paper and glue them to the white window. Encourage them to fill the entire surface. They may overlap pieces as this creates new forms and colors. The tearing creates soft edges rather than hard lines and gives a very nice effect. Just take care that the pieces are not too large, and that the children use small amounts of glue. You can control the colors offered to suit a particular theme or to insure a pleasant mix.

5. When they have filled the window(s) with color, write their names on the backs, fold back the two side supports and stand the triptych on the window sill so they can see the sunlight shine through. These are very nice when used carefully at home with a votive candle (the kind that comes in its own little glass holder) behind the center window. The children are captivated by the beautiful colored light.

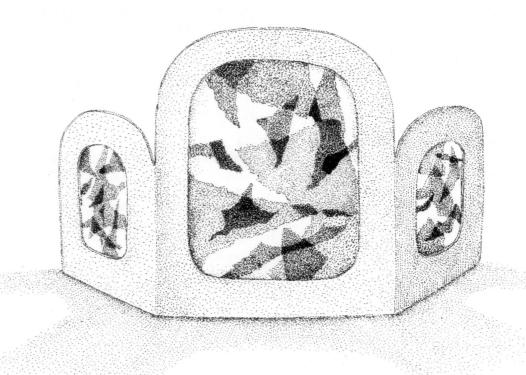

Folded Paper Boxes

These can be made in any size depending on how big a piece of paper you begin with. They are very handy for little gifts and treasures, and a set of three that nest one inside the other is lots of fun to open.

You will need

- heavy-weight paper—watercolor paper works best. Construction paper will also work, but the box just won't be as sturdy.

- scissors

- glue or tape

- optional—crayons or watercolor paints and brushes

What to do

1. If the paper is white, you may want to let the children color or paint it first. Watercolor painted boxes can be quite beautiful. Suggestions on how to work with watercolors are included in the Spring Whole Earth Home and Classroom section.

2. Take the paper and fold one corner up along the opposite edge. Trim away any excess. This gives you a perfect square to start (A). Now fold the other bottom corner up to the opposite corner. These two folds intersect at the center of the square (B). Establishing this center point is important for the other folds.

3. One at a time, place the tip of each corner of the square at the center point and make a crease. This gives you folds which look like a large square composed of four smaller squares in the center of your paper (C).

4. Again, using the tip of each corner, fold over to the opposite edge of the large square (D). Now almost the entire paper is made up of square folds.

5. Fold the tip of each corner just to the first folded line near it. This gives you a little triangular tip at each corner (E).

6. Find the central square. It will be composed of four smaller squares. This is the bottom of your box. Cut along the folded lines of two of the opposite sides toward the center, just to the edge of this center square (F).

7. Starting at the tips you did not cut, fold two opposite sides over and in until you reach the bottom center of the square. Leave these folded sides standing to make two sides of the box (G). Then fold the wing-like edges of the sides around so that

they overlap and make the other two sides.

8. Now finish the box. Fold the remaining two tips in, bringing them up and over the sides and pressing the tip down along the bottom inside of the box (H). You can put a drop of glue or a small loop of tape under each of these tips to hold them in place. Voilà! A box!

9. To make a lid, start with a square that is larger than the original one by about 1/32 inch on each side (in other words a tiny bit larger all around).

NOTE: Make sure to crease all your folds well—the box will be better for it. The children can help with the original decorating, but the folding needs to be pretty precise. To see the final box emerge is a wonder for them!

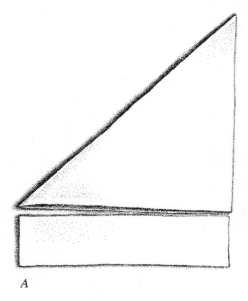

A

B

C

D

E

F

G

H

Candle Dipping

This activity gives the children a chance to see a candle take shape—something which is very magical to them. The candles make wonderful gifts. Although the activity may seem a bit complicated, it really is quite simple—you will see that once you've done it—and the results and experience are well worth it.

You will need

- candle wicking—available at craft and hobby stores. Don't be tempted to use string; the candles won't burn well. Get the appropriate wicking for one inch diameter candles.

- scissors

- masking tape and pen

- several pounds of pure beeswax—pure beeswax makes the most beautiful candles. You can mix it with less expensive paraffin if need be. Both are available at some craft stores. *

- towel, hammer and screwdriver

- two tall juice cans, cleaned and with tops removed

- a large pot—you'll need a pot big enough to hold the two juice cans in a water bath

- hot plate or stove

- newspaper

- potholders

- wax paper

What to do

1. Cut the wicking into pieces about 10-12 inches long.

2. Using a tab of masking tape, mark each wick at one end with the name of a child. This will prevent confusion after the candles are dipped.

3. The beeswax usually comes in large, thick chunks and needs to be broken into smaller pieces to fit in the cans for melting. I put the beeswax on a towel on the floor, and used a hammer and screwdriver to break it up by tapping the screwdriver with the hammer near the edge of a piece, driving it through the piece. The children love to help with the hammering and gathering of the pieces, so if you have the time, do this step with the children on the day before you want to dip candles. Tell the children that the fragrant wax is a gift from the bees!

4. About an hour or so before you want to make the candles, begin to melt the wax. Fill the juice cans 1/2 to 3/4 full with beeswax chips and pieces and place the cans in the pot almost filled with water. Experiment with

the amount of water to put in the pot, as you don't want the cans to float and tip over. A small, deep pot works better than a wide, shallow one.

NOTE: Take care to make this area **off limits** to the children and have an adult present **at all times**.

5. Keep adding chunks of wax until the liquid wax fills the can up to but no more than 3/4 full.

6. While the wax is melting, set up the area for dipping. Use a table that is the children's height. Cover one end of the table and the floor beneath it with newspaper. (Use newspaper for this activity. If you used oil cloth, it would get full of wax.)

7. When the wax in one can is all melted, carry it carefully to the table with potholders and place it near the end of the table. Seat yourself by the can and give the children their wicks. Have the children form a line, holding their wicks. One child at a time comes over to the can to dip. I held their hand and helped them dip their wick down into the wax singing, "Dip it down; pull it up." (The trick is not to leave the wick in the hot wax for too long. The wax then melts off the wick, instead of building up a new layer.) Straighten the wick for each child after they take it out by gently tugging on the bottom. As the candles thicken, you can also tap them on a piece of wax paper to flatten the bottom so they'll fit more easily in a candleholder. Then the children walk around the table and the room in a large circle (map out a

route beforehand) and go back to the end of the line to be in place for another turn to dip. This little walk gives the wax a chance to harden on the wick.

8. As the wax is used up, exchange cans with the fresh batch of melted wax in the water bath. Just add more chips to the old can and return it to the water bath for re-melting. (You want to keep the melted wax deep enough so the children's candles can be about 4-6 inches tall.)

9. Keep dipping until the children get tired or the candles are about one inch in diameter, whichever comes first. The youngest children may not make very thick candles, but you can dip them later if you want. The older children may get so involved that they don't want to stop, but don't let the candles get too fat.

10. The finished candles should be approximately 4-6 inches tall and one-half to one inch in diameter. Trim the wicks, wrap them in tissue paper and tie with a bow, or make a candleholder (see the activity, "Gift Making: Wooden Candleholders," on page 87) to go with them. What a "wonder full" thing to have done! This may seem complicated, but once you gather the materials and do it and see that it isn't so hard and is so much fun, you'll want to do it every year. Save the pot and cans of hardened wax; you can re-melt them next year. The children will look forward to it!

* Beeswax is available mail order from HearthSong. (See the Resource section at the end of the book.)

Pine Cone Fire Starters

These are simple and easy to make. They make wonderful gifts for anyone with a wood stove or fireplace, and they really work!

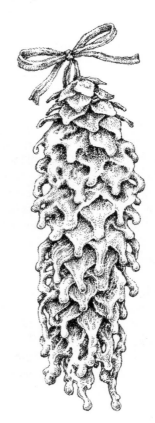

You will need

- wax—beeswax works fine but you may want to use the less expensive paraffin, or a mixture of both

- pine cones—any kind will do, although the longer, thinner ones work best

- yarn, ribbon or heavy thread

- scissors

- wax paper

What to do

1. Read the procedure for dipping candles, as this works much the same way (see the previous activity).

2. Melt the wax as you did for dipping candles.

3. Tie a string or piece of yarn or ribbon around the wide end of each pine cone. Make a loop about four inches long at the end of the pine cone.

4. Holding the end of the string, dip the pine cone into the melted wax and slowly bring it out. Hold it above the can for a few seconds until it stops dripping. Repeat the dipping once or twice. Then lay the wax-covered pine cone on a piece of wax paper to harden.

5. Tie the string into a bow at the top, or tie three or four pine cones into a bundle. Keep them at school until the gift giving time approaches. You can wrap them in tissue paper if you like.

INSTRUCTIONS FOR USE: Place the pine cone fire starters in the fireplace and light them like kindling to start a fire.

Rolling Candles

This is a less time consuming and easier way to make candles. The process is not as magical for the children as dipping, but the candles work, and if you have a group of only younger children, these may be more appropriate.

You will need

- wax sheets—available at craft and art supply stores. They look like a waffle or honeycomb and are made especially for rolling candles.*

- scissors

- wicking—available at craft and hobby stores

What to do

1. Cut the sheets into 8-inch squares and then into two triangles. Lay the wax triangle on the table with a straight edge toward you.

2. Place the wicking along the straight edge. Trim the wicking so that it extends one inch beyond the wax.

3. Begin to tightly roll the wax over and around the wicking, rolling toward the point of the triangle. The younger children may need help getting started.

4. Continue to roll the wax until the entire sheet is rolled. Generally, the more snugly you roll, the better the candle will burn—just don't roll so tightly that you crack the wax.

* Wax sheets are also available mail order from HearthSong or Nova. (See the Resources section for addresses.)

Modeling

Beeswax is just what it says it is—a truly natural modeling material. It is "wonder full" for young children as it warms in the hands and becomes more pliable as you use it. It's not cold to the touch and doesn't dry out like clay. It comes in colors, smells delightful and is very economical because it can be reused over and over and a little bit goes a long way. It comes in small slabs and is available mail order*, although you could ask your local toy or art supply store to carry it. They'll love it!

You will need

- slabs of modeling beeswax, cut into small pieces (approximately 3/4-1 inch squares)*

- small cake pan or baking tray

- basket or bag

What to do

1. If it is very cold weather, you may want to gently warm the beeswax by putting the little pieces in a small cake pan or baking tray and setting them near a heat source or in the sun.

2. Give each child one piece to start. Suggest that they put it in their little ovens (their closed hands) to warm and soften it.

3. Begin to manipulate the beeswax with your fingers, kneading, pulling, pushing and rubbing it. Stretch it out so thin that you can see the light shine through.

4. At first it's good to just get to know the material, not making anything in particular. Just play with it. You may want to do just this much several times.

5. Make various objects by warming the beeswax, working it to make it flexible and then shaping it into various forms, from birds to baskets to people. You might set up a little scene—some pine cones on a green cloth, a few rocks—and tell a story about the birds who live in this "forest." While you're telling—and after—the children can be making the birds, or whatever, to place in the forest. Send each bird home with its

maker, or re-soften the wax and form it into little beeswax patties to be used another time. Store in a basket or bag.

NOTE: You may also want to control the color selection at first, using only one color at a time, as this reduces color competition among the children and keeps the individual pieces a pure color. Eventually the older children may ask to use different colors to make hair, clothes, tree trunks, flowers, animals, etc. The younger ones should do quite well with one color at a time.

* Modeling beeswax is available from Meadowbrook Herb Garden and HearthSong (addresses in Resources section). They will sell it wholesale.

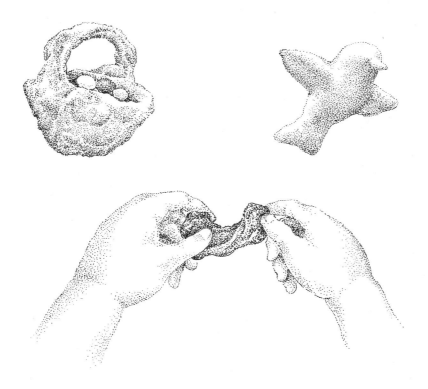

Decorating Candles

This is a way to make "store bought" candles much more special and beautiful. The decorating wax is very thin and goes a long way, so just give the children tiny bits at a time.

You will need

- candle decorating wax (very thin sheets of colored beeswax)*

- scissors

- trays or baskets

- candles—white or natural color work best

What to do

1. If you can't get special candle decorating wax, you can use regular modeling beeswax. Just use very tiny pieces and warm them in your fingertips until you can stretch them quite thin. Cut the sheets of decorating wax into small pieces (approximately 1/2 inch by 1/2 inch) and separate by color. Putting the different colors in individual trays, piles or baskets makes it easier to keep the colors separate while you are working.

2. Have the children take a small piece of wax, warm it in their hands and press it on the candle, smoothing it out and shaping the color as they go along. You can actually make a picture on the candle with the colored wax (the older children like to do this), or you can simply cover the candle with colors, which is easier for the younger children. Watch that each piece of colored wax is spread thinly and is firmly attached to the candle, not just stuck on. Also, the candle will burn better if the pieces of wax are added side by side, not one on top of the other.

NOTE: A lovely alternative decorating technique uses dried and pressed flowers, ferns, grasses, leaves—any flat natural object. After they have been pressed in a flower press or thick book, dip them in melted paraffin or use a small paintbrush to brush the melted wax onto the back side (if there is one) of the objects. Then press them gently and firmly onto the sides of the candle, brushing on more wax if needed. This method works better with larger-diameter candles rather than tapers.

* Candle-decorating beeswax is available from HearthSong and Nova. (See the Resources section at the end of the book.)

Valentine Hearts

This is fun to do and results in simple yet lovely Valentine greetings. Giving children the opportunity to make their own cards—rather than just buying them at the store—enables them to create rather than consume. Get out the colored paper, glue and scissors and let them have fun!

You will need

- colored construction paper—red, white and pink

- children's scissors (don't forget a pair or two for the lefties)

- white glue or glue sticks

- paper doilies—these aren't too expensive and add a nice touch

- red felt pen

What to do

1. Show the children how to cut hearts on a fold. Fold the paper, and along the fold draw an "elephant's ear." Cut this out on the fold, and you have a symmetrical heart! Vary the size of the elephant's ear to make larger or smaller hearts. This method is much easier than trying to draw the whole heart.

2. Glue different color and size hearts on top of each other, on doilies, etc. Let the children have fun.

3. The teacher can sit and cut lots of different size hearts for the little children to use. You can also be the message writer, if need be—perhaps using a red felt tip pen.

NOTE: This is also a great time to use saved scraps of fancy wrapping paper, foil papers, etc.

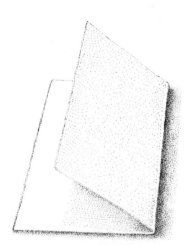

Valentine Mice

Hearts can also be mice!

You will need

- red paper hearts
- black felt pen
- bits of string about 3 inches long
- glue

What to do

1. Fold the paper heart in half and find the mouse. The nose and whiskers are at the tapered end— draw them in (see illustration).

2. For the tail, glue the string on the inside of the fold (opposite the end with the nose and whiskers), and let it hang out. There's the mouse!

3. Add a greeting on the inside.

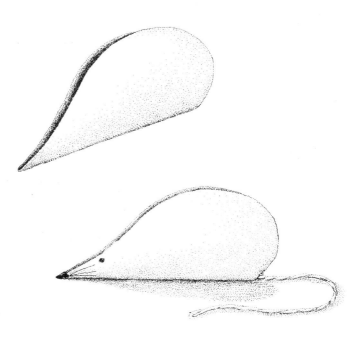

Valentine Swans

A bit more complicated to make, so these are better for the older children. They are very lovely.

You will need

- thick white paper or card stock (watercolor paper works well), or use white drawing paper for the swans and mount them on a piece of red construction paper

- pen or pencil

- scissors

- white glue or glue sticks

What to do

1. Fold the paper. Fold the swan pattern (see illustration), place it on the folded edge and trace it onto the paper. The older children may be able to do this.

2. Cut the swans out on the fold (you will be cutting 2 swans at once), helping those children who need it.

3. Open the folded swans. You can write a message inside the fold if you like. If you used lightweight paper for the swans, glue the swans onto red construction paper, fold the paper down the middle of the swans (like a card) and write the message on the inside of the card.

Growing Your Own Valentine

These need to be started a week to ten days before you want to send them home.

You will need

- cellulose sponges—they are much nicer to work with than the synthetic variety (get pink ones if you can)

- scissors

- small red construction paper hearts

- glue

- toothpicks

- marker

- a tray or dish to put the sponges in (they will be damp)

- grass seed—any quickly germinating kind

- mister or spray bottle

- white paper bowls or jar lids

What to do

1. Cut the sponges (one for each child) into heart shapes about 3-4 inches wide. You don't want them to be too small or too big. Write each child's name on a small construction paper heart, glue it to a toothpick and stick it in his or her sponge to keep track of whose is whose.

2. Have the children wet the sponges and place them on a tray or saucer.

3. Sprinkle the grass seed over the tops of the sponges, covering the surface.

4. Keep the heart gardens in a sunny window and water each day. The children love to do this, using a mister or spray bottle.

5. The grass should sprout quickly and grow tall and green. Wait until it gets at least an inch or more tall before sending them home.

6. Add more little hearts—one or two per garden. Glue the hearts to the top of the toothpicks and stick the other end into the sponge.

7. Send the heart gardens home on Valentine's Day. Use heavy weight white paper bowls or clean jar lids to support the garden on its way home.

NOTE: Another activity for Valentine's Day is to make sachets in the shape of hearts. See the winter craft activity, "Herbal Sachets," on page 88 of this chapter.

Chapter 3
Spring

S P R I N G

The Whole Earth Home and Classroom

**Cleaning House: Using Earth-Friendly
Products and Materials**

The idea of spring cleaning takes on new meaning for a teacher trying to make the classroom more earth-friendly, and for families trying to make environmentally sound changes at home. Examining the environmental impact of the basic categories of classroom and household items becomes essential to spring cleaning. The categories are cleaning and household supplies, including plastic and paper goods; arts and crafts materials; and food and drinks. The issues we need to consider when choosing these items for use in our homes and classrooms are toxicity—immediate and long term; packaging; and cost.

Toxicity—Immediate and Long Term

Whether a product is toxic when used is important to consider for supplies and materials that will be used around young children. But another aspect to consider is whether there are long term effects caused by the production or disposal of the product. Does its production or disposal cause the pollution of water, air or land? Is it made from petroleum or other resources that are non-renewable? How is it packaged? Is there excess packaging? How will it be disposed of?

It is important that we, as teachers and parents of the future generation, answer these questions and confront these issues. We need to apply this knowledge in our daily lives and in our classrooms. The children learn from what we do, whether we want them to or not. If we carry an attitude of conscious caring for the Earth, and act responsibly in choosing the most earth-friendly products available, this will establish good working habits and have a lasting influence on the children. The other thing to remember is that once we get used to choosing and using more earth-friendly products, it will just become second nature to do it. The key here is a change in attitude.

Fortunately, the task of researching the background of supplies and materials is being made easier as more people become concerned about the effects of these products on their own health and the health of the Earth. Many books are now available to help us make sound choices from the myriad of products available, for example Debra Lynn Dadd's three books, *Non-Toxic, Natural, and Earthwise: How to Protect Yourself and Your Family From Harmful Products and Live in Harmony With the Earth, Non-Toxic & Natural: How to Avoid Dangerous Everyday Products and Buy or Make Safe Ones*, and *The Non-Toxic Home: Protecting Yourself and Your Family From Everyday Toxic and Health Hazards*, and her newsletter *The Earthwise Consumer.*

Others are *Clean & Green: The Complete Guide to Non-Toxic and Environmentally Safe Housekeeping* by Annie Berthold-Bond, *The Green Consumer* by Hailes and Makower, and their newsletter, *The Green Consumer*, as well as *The Green Pages: Your Everyday Shopping Guide to Environmentally Safe Products* by Steven J. Bennet. (The publisher's names and addresses are listed in the Resources section.)

Packaging

Another factor to consider when buying supplies and materials of all kinds is the packaging. Work toward the idea that less is more in this realm. That is, buy things in large containers so less packaging is required for each product. Single serve items or individual glue jars, etc., should be avoided. If you need paste or glue for each child, buy a large jar and put some on saucers or in bowls for individual use. Whenever possible, choose products which can be recycled in your area. Glass is endlessly recyclable, that is, it can be made into new glass over and over. The same is true of aluminum. The technology for and the ability to recycle plastics is growing, but not all kinds are recycled in all areas. And, remember that plastics are petroleum based products, that is, they are formulated from a non-renewable resources. Although it is very difficult to find certain kinds of items in recyclable containers, just keep looking and asking for alternatives and, in the meantime, make the most environmentally sound choices you can. And buy as little packaging as possible.

Cost

This is a tricky one as environmentally sound alternatives may cost more initially. But, if you were to factor in the hidden costs resulting from production, use and disposal, such as damage to the environment and health problems resulting from pollution, the picture changes. For example, buying cloth napkins or placemats might cost more initially than paper napkins, but you'll have them for years. When they are no longer good enough for table use, they can be used to mop up spills, etc. That will save more than a few trees over time. Another example is crayons made of beeswax—a renewable resource—as opposed to crayons made from petroleum products. The beeswax crayons cost more initially, but they don't break and are more pleasant to use. If you clean them, they can last for years, depending on use. Beeswax crayons are available from HearthSong and Nova (addresses in Resources section).

A good guideline is not to make choices based solely on money. Make your choice by considering the question from an environmental perspective as well. If you need to buy fewer crayons because they cost more, so be it. You can teach the children to share and to really appreciate and care for the things they do have.

In short, if you use it once and throw it away, try not to buy it at all and consider responsible alternatives. Check with your local licensing agency regarding health regulations. Ask your school to order bulk paper supplies from a supplier who provides recycled paper products, such as paper towels, if you must use them, and toilet tissue. And teach the children by example and gentle reminders to use as little of these disposal items as possible, for example, just one paper towel to dry their hands.

Now that we have some guidelines for choosing supplies and materials, let's take a look at the different types we use.

Cleaning and Household Supplies

Here is another area where less can be more. Plain soap, vinegar, baking soda and borax can accomplish most of your cleaning needs—even taking care of germs, which can be a real day care nemesis. In her book, *Nontoxic, Natural, and Earthwise,* Debra Lynn Dadd tells of a hospital that experimented for a year with a borax solution as a disinfectant (1/2 cup of borax to 1 gallon of water). It met all their germicidal requirements and was an excellent deodorizer as well! Using simple, non-toxic cleaning products is not only better for the environment, it's much less expensive. The book *Clean & Green* by Annie Berthold-Bond is subtitled, "485 ways to clean, polish, disinfect, deodorize, launder, remove stains—even wash your car without harming yourself or the environment." She offers many "recipes" for making your own cleaners for common household products.

Schools are often faced with insect or rodent problems of various kinds and again, there are non-toxic alternatives. Encourage your school to explore these for large scale use, and ask that your classroom be spared any applications of toxic materials. Be sure to keep things clean and dry to remove sources of food and drink and, thus, encourage critters to look elsewhere. If you have a problem with mice consider a live trap (such as Havahart), which lets you relocate the mouse to a more suitable habitat. Again, Debra Lynn Dadd's book offers many "earthwise" alternatives.

Arts and Crafts Materials

The key here is to stay, as much as possible, with water-based materials (paints, pastes, glue, markers), as they are generally non-toxic. Read labels and ask questions. In general, stay away from professional artists' materials. They are often more toxic. If something has a strong smell, like some permanent magic markers and rubber cement, it is best to look for less toxic alternatives. The Arts and Crafts Institute (715 Boyleston Street, Boston, MA 02116, 1-617-266-6800) evaluates art materials and labels them with seals as either AP (Approved Product) or CP (Certified Product). They will send you a list of non-toxic art supplies. Send a self-addressed stamped envelope.

Crayons made from beeswax (available from HearthSong and Nova) are available in blocks as well as sticks. The block crayons are very nice for preschool children, as they are easier to hold and they don't break.

Use recycled paper for drawing and crafts whenever possible, and encourage the children to use both sides of their paper. Look for natural paints, as often standard tempera, watercolors and finger paints are formulated with various chemicals, including formaldehyde. Natural paints are available by mail. I used the Stockmar watercolors (also available from HearthSong and Nova) which are available in a wide array of colors. My jars of red, blue and yellow seemed to last forever. Mix them in glass jars with tight fitting lids and store the remaining paint in the jars for reuse. Mix just a bit of the color (start with a teaspoon) in a 1/2 quart of water and test it for strength. Remember that watercolors are not supposed to be as opaque as tempera paints. If you buy just the three primary colors—red, yellow and blue—the children can mix any other colors they need right on their paper. This can be a very exciting exploration. If you dampen the watercolor painting paper first by dipping it in a tray of water and then gently wiping it with a clean damp sponge, the colors are more fluid on the paper and the painting experience is delightful. Watercolors are available from:

HearthSong
P.O. Box B
Sebastopol, CA 95473
1-800-325-2502

Nova
27 Eagle Street
Spring Valley, NY 10977
1-914-3757

Food and Drink

If we want to be more conscious about the things we have around the children, it follows that we would also want to be aware of what goes into the children, as well. Make an effort to obtain the most healthful, wholesome snacks and meals possible. Ideally, this means giving them food that is free of chemical additives and preservatives and that is organically or bio-dynamically grown whenever possible. Food that is produced in this way is not only better for the children but better for the Earth because the farming methods used to produce this food are life sustaining. Again, organic whole foods may be more expensive, but it would be better to have less of something that is really good and consume conservatively, than to have lots of mediocre food that can actually encourage wastefulness. At snack and meal times we had a rule that everyone would join us at the table and have, at least, a taste or tiny portion. This encourages the children to stretch their taste buds and contributes to an enjoyable social atmosphere at meal times.

A short verse of thanksgiving helps to set the mood before eating. I use this one:

> Earth who gave us all this food,
> Sun who made it ripe and good,
> Dearest Earth and dearest Sun,
> We will not forget what you have done.
> —*Christian Morenstern*

Organically grown produce is becoming more readily available, but you may need to track down a source in your area. Check the local natural foods store or ask at the agricultural extension service in your area for the names of farmers who use organic methods. Make use of your own garden as well; then you can be totally sure of the freshness and the quality of what you consume. You can also get organic dry goods such as grains, flours, nuts, and seeds mail order from Walnut Acres in Penns Creek, Pennsylvania. Call 1-800-433-3998 for a catalog.

Good food, just like good crayons and good paper, should be thoroughly enjoyed and appreciated. Take the time to foster appreciation and gratitude and habits of careful, conservative use in the children while they are young, and it will bear fruit for years to come.

Bringing Nature In: The Season's Garden in Spring

For more detailed information about setting up a Season's Garden, see the section on " Creating a Seasonal Garden" in the fall chapter.

Colors: Pastels—pale pinks, light blues, spring greens, yellows.

Objects: Fresh flowers, small bird nests, dyed eggs (as a symbol of rebirth), polished stones or special rocks, budding branches (in water).

Wreath: Remove the evergreens from your winter wreath. Bind it with wild grasses or weeds, adding flowers here and there for a touch of color. You could also twine ivy around the straw base, again adding flowers for color.

Plantings: Just as the children create small dish gardens for a springtime activity (see "Dish Gardens"), have them help you create a spring class garden in a large clay plant saucer. If you have one that held moss for the winter, transplant the moss to the outdoors, or keep the moss in one area of the saucer and place soil in the remaining area, being careful not to disturb any small bulbs you may have planted in fall. Sink a large sea shell, jar lid or tiny clay saucer down into the soil so that its edges are even with the top of the soil. This can become a small pond when filled with water. Lightly dampen the soil with a plant mister, and have the children help you sow grass seed all over the saucer garden. Remember not to sow the seeds too thickly, but make sure all the soil is covered with seed. Sprinkle with a light "blanket" of soil and thoroughly mist again. Remember to have the children mist the seeds or grass each day.

You can add a small blooming branch or flowers either directly into the soil or in a small vase of water, and the children can help you create lots of little beeswax characters to live in the garden. A small white duck or green frog can live in the little pond; perhaps a rabbit or deer in the tall grass; maybe a little red gnome by the rocky, mossy area. A bluebird might build his nest in the budding branch; and there might even be eggs in the nest. The Season's Garden can become a whole story. The possibilities are endless! For more detailed information on working with beeswax, see the winter activity, "Beeswax: Modeling" on page 102.

When the grass grows too long, you can cut it using a pair of scissors. We used to send our fresh grass clippings home as a treat for one of the children's pet rabbits, but you could also put them on your window sill for nest-building birds to borrow or, at least, put them in your compost pile.

Round Wind Wands

These "wind wands" or "round wind catchers" are lovely for those first warm spring days.

You will need

- cane—one piece per child, approximately 36 inches long

- masking tape

- streamers—the crepe paper variety, approximately 1-1 1/4 inches wide and 18-24 inches long in springtime colors

- transparent tape

What to do

1. Start with a piece of cane about 36 inches long. If the cane is thin, make it more sturdy by holding both ends together and twisting it so that it wraps around itself.

2. Now bend the (twisted) cane into a round shape, and thoroughly secure the ends with masking tape. You want to have a circle that is at least 5 inches in diameter.

3. Have the children select five to seven streamers.

4. Attach the streamers side by side by folding a streamer end over the cane and taping it both to itself and to the caning.

NOTE: These round wands are good for younger children because they can't poke anyone accidentally.

5. Bring the wind wands outside and "catch" the wind with them. Some children may want to run with the wind to catch even more wind.

Streamers on a Stick

A magic wand with frills—these are very appealing to the children and provide lots of possibilities for imaginative play.

You will need

- streamers—ribbons or the crepe paper variety work well—cut to the width desired. The streamers should be 1 to 1 1/4 inches wide and 24 inches long.

- sticks or dowels approximately 12-15 inches long

- transparent tape

What to do

1. Let the children choose the colored streamers for their wand—three to five per wand.

2. Twist the streamers tightly together at one end, and then attach them to one end of the stick with the tape.

Make sure that the tape is adhering to the stick and not just to the streamers. By twisting the streamers together and placing them on the top of the stick, the tape has lots of contact with the stick. If you are using regular sticks, wad some tape over the top end to make them less "poke-y."

3. Depending on your group, you may decide not to allow them to be used indoors, as the space may be too confined. Make some rules about the use of wands before you take them outdoors. Make sure the wands have lots of room to move, and have the children be careful if you allow them to run, so that they do not bump or poke others.

Blowing Bubbles

Make-your-own bubbles are pleasant to use and much more economical than commercially-made bubbles. Keep a quart jar full of bubble mixture handy, and whenever you can't go outside on a rainy day, sit together and blow bubbles during outdoor playtime. It is magical—fun, quiet and soothing—something to look forward to on rainy days. Beautiful rainbow bubbles—what a joy to experience them—indoors and out!

You will need

- mild dish soap—certainly something non-toxic
- glycerin—available at pharmacies
- tablespoon
- quart jar
- spoon
- straws or bubble wands
- little jars or cups to hold each child's portion of bubble stuff

What to do

1. Mix bubble stuff by gently combining 2 tablespoons of liquid dish soap and 2 tablespoons of glycerin with a quart of water. Stir them together gently so the mixture isn't too bubbly at this point.

2. Make the rules clear before the children begin to blow bubbles. When we blew bubbles on rainy days, we all sat around a large table. We had the following three rules:

Rule # 1—Only pop your own bubbles—not someone else's.

Rule #2—Don't blow into the jar, as this quickly exhausts the bubble stuff.

Rule #3—Stay in your seats—no running around chasing bubbles indoors. Watch them float around the room.

For indoor bubble blowing, these rules helped to create a quiet, calm atmosphere. Outdoors, of course, is different, as the children can chase the bubbles. Just don't run with the bubble stuff!

Tips on Containers: Glass baby food jars with the labels removed (also used for painting) can be washed and reused. The glass isn't a problem if you are sitting down indoors and things don't get wild. Use them outdoors only if the children aren't walking or running with their containers.

Tips on Blowers and on Blowing:
Straws work well for indoor blowing
and are available in large quantities. Be
sure that the children dip the straw into
the bubble mixture, take it out and blow
the bubbles gently into the air. If they
blow directly into the jar, the bubble
stuff gets too bubbly and is quickly
used up. Tell the children that their bubble stuff won't work if they blow into
the jar (this is true—too much air gets
into the liquid and it won't blow single
bubbles). However, since it does make
a pretty fountain of bubbles, allow this
at the end of bubble time.

In addition, some children need to learn
to blow—watch the little ones who
may only know how to suck in! Hold
their hand by your mouth and blow gently on it. If the little ones still can't do
it, sit them near you and blow their bubbles for them. The older children need
to learn to blow gently (again, demon-
strate on their hand), as blowing too
hard pops the bubble just as it comes
out.

Another discovery the children made is
that if you're sitting at a table, you can
blow bubbles on the table. Dip your
straw into the bubble stuff, and while
holding it near the table, gently blow
the bubbles onto the table top, creating
"bubble domes." The children had a
great time making connecting bubble
villages and blowing bubbles inside
bubbles.

For outdoor blowing, bubble wands
work better, and you won't have to
worry about the children walking or
running with straws in their mouths.
Collect wands over time; ask each child
to bring one from home or make them
by twisting wire into the desired shape.

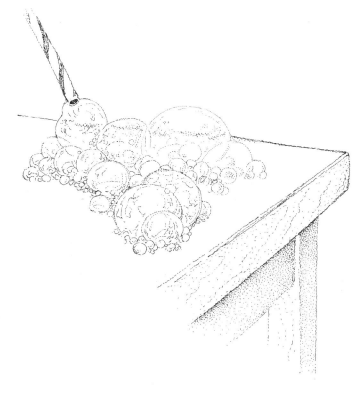

Pinwheels

Perfect for a breezy day. See what the wind can do! Or the children can create a wind of their own (if nature isn't cooperating) by running or blowing!

You will need

- colored construction paper or heavy white watercolor paper that the children decorate with crayons

- scissors

- straight pins

- new, unsharpened pencils with erasers

- scissors

- optional—scraps of paper, tape

What to do

1. Cut the paper into 7-inch squares. You can vary the size with a larger square, but the pinwheel will be more floppy. Don't make it bigger than 8 1/2 inches.

2. Determine the center of the square (see illustration). This is done by lightly folding tip number one to tip number three and tip number four to tip number two. You don't need a strong crease. The center is where the two folds intersect.

3. Now, using the scissors, cut in on each fold line about 3/4 of the way to the center. Leave the last inch of each fold uncut.

4. With a pin and pencil ready, fold every other tip (you now have eight) into the center and overlap them. Push the pin through these overlapping tips, through the center of the pinwheel and into the eraser. Don't let the pin stick out the other side of the eraser.

NOTE: A little paper washer reinforces the center of the pinwheel. Cut a small circle of colored paper (about 1-1 1/2 inches in diameter). This doesn't have to be a perfect circle— just do it freehand. Put a tape loop on the back of the circle and press it onto the overlapping tips of paper at the pinwheel's center. Then insert the pin. This provides extra support at the stress point.

5. Show the children how to make the pinwheel turn by blowing on it. Then let them take the pinwheels outdoors and see what the wind can do.

Kites

Kites for young children do not need to be the elaborate kind which are totally sky worthy. Though these are nice, and all children should have the experience of flying a kite with an adult, the children are often quite happy to make a simple little kite of their own that will flutter up behind them as they run. Then they aren't disappointed if it's not a windy day!

You will need

- heavy colored paper—construction paper or watercolor paper, card stock or posterboard

- scissors

- crayons

- pen

- tissue paper or crepe paper—cut into streamers approximately 2 feet long by 1 inch wide

- clear tape

- heavy string

- sharp pencil

What to do

1. Cut a standard diamond kite shape from paper. It should be approximately 12 inches from top to bottom and 9 inches from side to side. The size can vary, as long as it's not too small. If you trace the form on paper, older children can cut it out by themselves, depending on the thickness of the paper.

2. If you are using white paper, have the children decorate their kites with crayons. Put their name in a corner.

NOTE: You can also have the children paint their kites with watercolors. See the description of watercolor painting in the Whole Earth Home and Classroom section of this chapter. They're beautiful!

3. Let them choose three to four streamers and attach them firmly to the back bottom corner with tape.

4. Attach a length of string (approximately 1 yard) to the center of the kite by punching two holes (a sharp pencil will do fine) just to the left and right of the center of the kite and threading the string through the holes. Tie the string in a knot on the underside of the kite, and then reinforce the holes by placing two or three pieces of clear tape over them on both front and back.

5. Once outdoors, let the children run with their kites trailing behind them. It's amazing how much they enjoy these kites with their fluttering streamers.

Dish Gardens

A little saucer can become a springtime wonderland!

You will need

- small clay saucers—approximately 4 inches in diameter, the kind you put under flower pots

- crayons

- potting soil

- large spoons

- grass seed—a quickly sprouting variety if possible—you don't need much. Ask parents for leftovers.

- bowls

- plant mister or spray bottle

- small sea shells

- beeswax

- tiny branches

What to do

1. Often nurseries will sell clay saucers to you at a discount if you buy in quantity and tell them they are for a school. Or use small bowls (each child could bring one from home) or even the heavy paperboard "china" bowls sold in grocery stores. They just need to be low and wide.

2. Write each child's name on the bottom of the saucer with a crayon.

3. Let the children decorate the outside rim of their saucer with crayons.

4. Set up a potting table with bowls of soil, large spoons, a bowl of seed and a plant mister. The children spoon the soil into their dish, filling it about 3/4 full. They often prefer to use their hands for this. Then moisten the soil with the mister, sprinkle on a good layer of seed (not too thick, but make sure to cover the soil surface well) and cover with a thin layer of soil. This last thin layer is not absolutely necessary, but the children like to "tuck the seeds into bed." A final watering with the mister, and the garden can be put on a sunny window sill to be watched with anticipation for the first signs of life.

NOTE: Moisture is the key to sprouting. The gardens should be watered once per day (more if they seem to be drying out—perhaps give them a good drink before you leave for the day). The mister or squirt bottle works well for this as it doesn't flood the seeds and dislodge them. Most of the children will come eagerly to do this task each day.

5. Have the children add a little upside-down shell to the center of the garden before the grass starts to sprout. This tiny basin will hold water and becomes a little pond in the garden.

6. Once the grass has sprouted, make the gardens a lively place. Use modeling beeswax to embellish the gardens. The children can form all sorts of little things from it: rabbits, birds, baskets of colored eggs, flowers, etc. For a more detailed description of using the beeswax see "Beeswax: Modeling" in the Winter chapter.

7. Push a tiny spreading branch (the tip of a large branch) into the soil to become a little tree. Glue on tiny balls of pink tissue paper for blossoms and, perhaps, add a beeswax bird sitting on a beeswax nest. Fill the shell pond with water and add a beeswax duck sitting on the edge. The possibilities for these gardens seem endless. Use your imagination—the children will help you with this.

8. A little sprig of forsythia or other blossoms add a nice touch on the day the gardens are going home.

Butterfly Pop-Up Cards

This is a simple idea that is very adaptable for any occasion. The pop-up makes it intriguing. You can make other little pop-up figures, depending on the occasion and the interests of your group—birds, rainbows, hearts. Older children will want to design their own. Adjust the size of the paper figure to the size of your card, and once the older children get the hang of the accordion fold, be prepared for them to want to fold everything in sight!

You will need

- card paper—card stock is very special, but construction paper will do—choose light, springtime colors or have the children color or watercolor paint white paper

- crayons

- construction paper—some cut into little butterflies (see pattern) and some cut into little strips 3/4 inch wide by 6-8 inches long

- scissors

- white glue or glue sticks

What to do

1. Fold the card paper in half or in quarters to make the size card you want. The size of the butterfly should also vary with the size of the card. The size butterfly illustrated works with an 8 1/2 x 11 inch sheet of paper folded in quarters (in half top to bottom and in half again side to side).

2. Let the children color the front (and maybe the back) cover of the card. If you are using construction paper, be sure to choose lighter colors so the drawing and decorating will show.

3. Give each child a butterfly and a small strip of paper. Show the children how to fold the strip accordion style—front to back, not over and over. This becomes a little paper hinge.

4. Put a small dot of glue on the underside of the butterfly and attach it to the top of the hinge.

5. Glue the bottom of the hinge to the center of the right-hand side of the inside of the card.

6. When you open the card, the butterfly will gently rise up.

7. Make sure the children's names are on their cards. Write a message if desired, or allow the children to decorate the inside of the card with crayons.

NOTE: You could also have the children do a watercolor painting from which you cut the part that will become the folded card and the small butterfly. Add lots of water to your colors so you get pale, pastel shades.

Tissue Paper Butterflies and Mobiles

These are simple to make, yet very beautiful. The children will love them and may want to make several.

You will need

- colored tissue paper cut into rectangles 3 x 5 inches, with rounded corners and in light spring colors

- pipe cleaners—white is nice, but any color will do—approximately 5-6 inches long

- thin string, heavy thread or embroidery floss

- scissors

- sticks or dowels

- pieces of cane

- ribbons or crepe paper

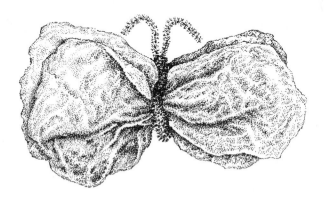

What to do

1. Take two tissue paper rectangles and place one on top of the other. Using two different colors can be nice.

2. Gather them in the center and bind them with a pipe cleaner by twisting it several times around itself. The ends of the pipe cleaner will be the butterfly antennae; the twisted section will be the body.

3. Separate and fluff up the tissue paper wings and bend back the pipe cleaner antennae to shape your butterfly.

4. Tie a heavy thread or light string through the top of the butterfly (go under the top of the pipe cleaner). By holding this string and moving their arms around, the children can make the butterflies fly. Or tie the butterfly's string on the end of a stick. This makes it very special, like a magic wand. Just watch that no one gets poked and that the children aren't running with the sticks—running is for "stickless" butterflies only.

5. A tall vase full of colorful butterflies on sticks or dowels makes a lovely springtime centerpiece.

6. You could also make a lovely butter-fly mobile. Bend the caning into a ring (approximately 6-8 inches in diameter) and bind it with heavy thread, string or embroidery floss. Then wrap it with colorful ribbon or crepe paper. Tie the butterflies onto the ring of cane, varying the lengths of string so that they hang at different heights.

7. Using three equal lengths of heavy thread, string or embroidery floss, attach each length to the rim of the ring. Space them equally around the ring so it will be balanced. Gather the ends of the strings together and knot. Hang the mobile over your Season's Garden. Have each child make a mobile, and they will enjoy taking them home to share with their families.

 BRINGING NATURE IN: SPRING CRAFTS

Natural Egg Dyeing

The egg is a wonderful springtime symbol of rebirth and the new growth which begins at this time of year. Using spices and vegetable scraps to color eggs is magical, fun and inexpensive. A bit of alchemy in the kindergarten!

You will need

- hard-boiled eggs

- raw eggs

- onion skins from yellow onions—the paper-thin, outer brown skins. (Ask your grocer if you can clean out the onion bin.) You will need approximately 3-4 cups of skins. These make a beautiful golden to brown color dye.

- turmeric, the spice. Tie up a heaping tablespoon or two in a bit of cheese-cloth to make it less messy and to avoid speckles (although speckles are nice, too).

- red cabbage—the large outer leaves work best

- rubber bands

- vinegar

- measuring spoons

- white eggs

- large pot(s)

- stove or hot plate

- egg cartons

- long-handled spoon

- tiny leaves or flowers

- pieces of nylon stocking or cheese-cloth

- twist ties

What to do

1. For onion skins and turmeric, boil the material in a one- to two-quart pot of water for 30 minutes or so. Then strain and add warm, hard-boiled eggs. You can also boil them in the dye pot instead of in advance, but be aware that some will crack. Add two tablespoons of vinegar when you add the eggs. Simmer the eggs for 10-20 minutes and remove them with a long-handled spoon. Cool them in an egg carton.

2. For red cabbage, wrap the **raw** eggs in cabbage leaves, completely covering the eggs. Hold the whole bundle together with rubber bands. Add 2 tablespoons of vinegar and the egg bundles to a pot of water and simmer just below boiling for 10-20 minutes. Remove them from the pot and unwrap when cool. This makes lovely blue eggs with a tie dye effect.

NOTE: Raw eggs seem to take the color better than cooked when using this method.

3. Resist dyeing—this method works especially well with the onion skins or turmeric. Have the children collect tiny leaves or flowers. Press them flat onto the eggshell and hold them in place with pieces of recycled nylon stocking or reusable cheese-cloth. Pull the material tightly around the egg and hold it together at the bottom with a twist tie. Dye as normal, and when you remove the "wrapper," your little leaf or flower will have left its impression.

4. Experiment with other fruits, vegetables or plants. This is not an exact science, so be creative, have fun and expect some unexciting as well as exciting results.

May Baskets

Some people celebrate the first of May and the coming of flower-filled warmer weather by dancing around a Maypole. Another way to celebrate is by making little cone-shaped baskets which you quietly, and in secret, hang on your neighbor's doorknobs, either in school or at home. Your tiny gift will bring lots of happiness. Surprises like this are always nice, and the children love to be the "surprisers."

You will need

- 8 1/2 x 11-inch sheets of white construction or watercolor
- paper
- scissors
- crayons
- clear tape
- small flowers and greens—clover blossoms, dandelions, forget-me-nots, crocus, buttercups, violets, etc. Keep them in water until you're ready to fill the baskets.
- paper towels or tissues and aluminum foil

What to do

1. Cut the 8 1/2 x 11-inch paper into four equal rectangles. Fold the original sheet in half side to side and then top to bottom. The rectangles will measure approximately 4 1/4 x 5 1/2 inches. Each of these will become a basket.

2. Cut handles 1/2 x 11 inches by cutting 1/2 inch wide strips from 8 1/2 x 11 inch sheets of paper.

3. Have the children color the baskets and handles with crayons or paint with water colors, or use colored construction paper in light spring colors.

4. Shape the rectangles into cones as follows (see illustration). Wrap corner B down and overlap it over corner C about 1/2 inch, keeping the edges straight. Tape these two together. Corner D becomes the bottom of the cone. Wrap it around to form a point and tape it in place.

5. Add a 1/2 inch wide paper handle by taping it to the inside top of each cone. Put the tape over each end of the handle and press it into place inside the cone.

6. Let the children choose a few flowers for their baskets. It's nice to bind the flower bottoms with a bit of wet paper toweling or tissues and aluminum foil. The flowers will stay perky longer.

7. Quietly go through the school, placing the baskets on classroom doorknobs. If you can choose a time to do it undetected, all the better. If the children will be taking their baskets home for a neighborhood surprise, perhaps you can make these the day before May Day, and tell the children how they can quietly hang the baskets on a neighbor's door early in the morning before coming to school.

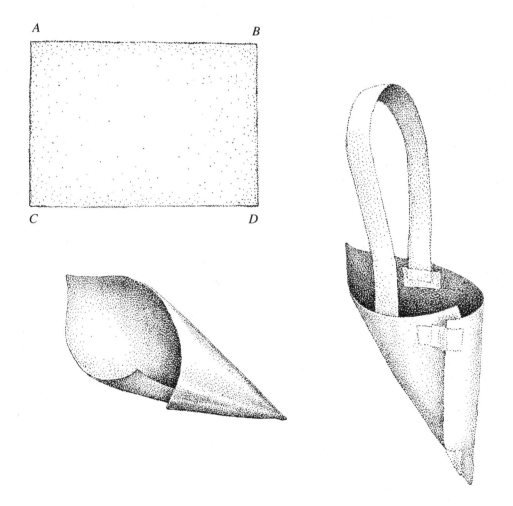

Flower or Leafy Crowns

These make beautiful springtime costumes. We made extra ones to keep in the classroom. The children wore them out! They are also wonderful for all to wear to a May celebration as you dance around the Maypole. This is another one of those handwork projects that can be kept in a basket and done over a period of days. I required everyone to make one, but they didn't have to work on it every day. Some children will sit down and do the whole thing in a day, others need coaxing to come and reminders to finish. The little ones need lots of assistance which sometimes the older, more capable ones can provide as well as the teacher. The Leafy Crowns originated one year for the benefit of some older boys (6+) who just couldn't bear the thought of wearing a flower crown. It somehow provided them with the out they needed and still allowed them to participate. They wore the Leafy Crowns to the Maypole celebration.

You will need

- ribbons—choose light spring colors: pink, light green, light blue, pale yellow. I used 3/4 inch wide grosgrain ribbon. It isn't as slippery as the shiny varieties and is easier to work with.

- tissue paper—again choose light spring colors—although I did use a deeper green and a dark green-blue for Leafy Crowns. Cut the tissue paper into rectangles 2 inches wide by 5 inches long and sort according to color.

- needles and thread

What to do

1. Give each child a ribbon cut to fit his or her head (don't forget to add another 8-12 inches for the knot and streamers). Also, give each child a needle knotted and threaded with about 12 inches of doubled thread.

2. Bring the threaded needle up through the ribbon about 7 inches in from the end to allow space for knotting and streamers. This also puts the knot on the back side of the ribbon.

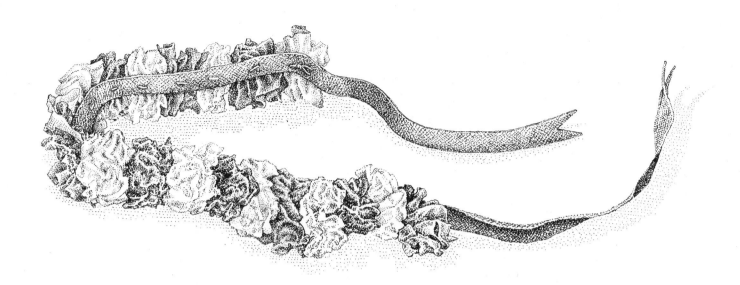

3. Have the children choose a rectangle of colored paper and sew a running stitch lengthwise down the center of it, scrunching it as they go. Then gather this "blossom" or "leaf" by pushing the tissue paper down to the end of the thread. Tack the blossom onto the ribbon by sewing down through the blossom and the ribbon, and then up through the ribbon again, so the needle is now on top of the ribbon, ready to stitch another tissue paper blossom.

NOTE: Help the children fluff and/or scrunch the tissue paper to make it more "blossomy."

4. Continue to gather and stitch these blossoms onto the ribbon until you reach a point seven inches in from the other end of the ribbon. After stitching on the last blossom, knot off the thread on the back of the ribbon. The closer the blossoms, the more beautiful the crown will be. Choose a random selection of colors, or encourage the children to set up a pattern and stick with it—pink, green, yellow; pink, green, yellow or dark blue, dark green; dark blue, dark green. The children enjoy this sequencing, anticipating what will come next.

5. When the ribbon is full of blossoms, except for the knot and the streamer allowance at each end, tie it around the maker's head and marvel at what a beautiful thing they've created!

Pressed Flower Cards

This project needs to be started a couple of weeks before you want to finish it so that the flowers have time to dry. The cards make lovely Mother's Day greetings. (See also "Butterfly Pop-Up Cards" in this chapter.)

You will need

- fresh flowers, grasses, weeds, etc.

- a flower press or heavy book, e.g., a telephone book

- some absorbent paper, such as plain newsprint or construction paper

- nice paper for the cards—construction paper (in pretty spring colors: green, yellow, pink or light blue) or water-color paper

- glue sticks, or white glue, small dishes and cotton swabs

- wax paper

- crayons

- colored pencils

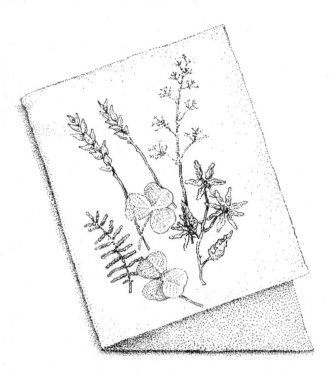

What to do

1. A couple of weeks before you want to make the cards, gather flowers, grasses, weeds, etc., to dry. You can dry anything, even weeds look lovely. Keep in mind that the thicker the item, the less likely it will dry well. For example, a dandelion is too fat to dry well, although you can certainly try one as an experiment.

2. Begin pressing the flowers, etc., before they wilt, so bring the flower press (or phone book) outdoors, or bring the items indoors fairly soon after you've picked them. Arrange them on the blotter sheets of the press or between the folded sheets of absorbent paper if you are using a heavy book. The pressed flowers dry the way they are placed in the press, so take time to straighten them out and spread the flower petals. The older children love to help with this. Also, place each item separately on the paper—do not overlap them. If using a phone book, do not use pages next to each other. It is better to leave chunks of pages between each place with flowers.

3. Tighten the fastening screws or belts of the press, or place something heavy on top of the phone book—another phone book will do.

4. Carefully check the drying items after several days. When they are ready to use they will actually be dry. Then carefully remove the flowers and store them flat—perhaps between sheets of paper—until you are ready to use them. You can then add other items to dry. Change the absorbent paper when needed.

NOTE: Drying things from nature is something you can do year round, starting with colorful leaves and fall flowers, winter grasses and hardy weeds, and springtime buttercups and clover leaves. The technique is the same, the children love to do it, and you can build up an accumulation of items to be used for cards, tables, decorations, posters or bulletin boards or to put in the Season's Garden.

5. Make cards by folding a sheet of paper in half. Paper 8 1/2 x 11 inches works well, but use any size you like. Just don't make it too big or you will use up all the dried flowers very quickly.

6. On the front of the card, the children arrange a few of the dried items, then glue them down. Glue sticks are less messy to use, but a dish of white glue with cotton swabs (and reminders to use just a little!) will work. Apply the glue lightly to the paper and gently press the flowers onto it.

7. Lay the card between a folded sheet of wax paper so it won't stick, and put it in or under a heavy book for 30 minutes or so to give the glue time to dry.

8. The children can draw a picture inside the card. You can write a short message for the children using colored pencils rather than magic markers. They have a softer, lighter touch.

Flower Necklace or Lei

These are beautiful and can be made with any number of different flowers; experiment and be creative. Hyacinths make a particularly lovely, sturdy, sweet smelling version. (Other blossoms may be used, just look for those that are small and sturdy, something that can be strung easily and will lie nicely.) They are excellent Mother's Day gifts and very nice to wear when dancing around the Maypole.

You will need

- fresh flower blossoms—one good-sized hyacinth will make one lei

- needles

- thread—start with a 24-inch length of doubled thread for each child

- tape and pen

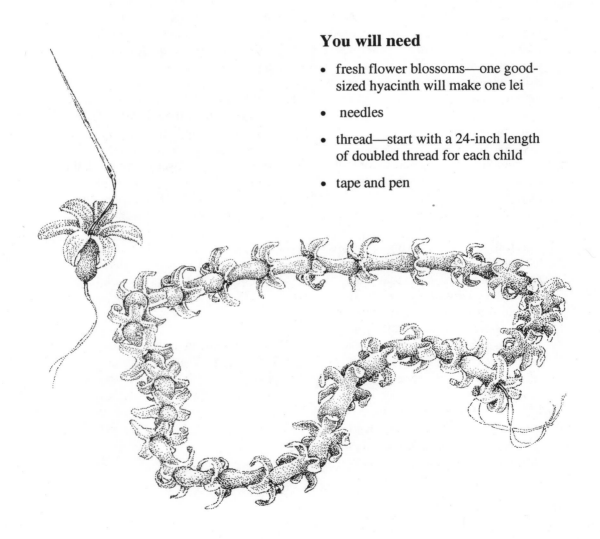

What to do

1. Gather flowers. You could ask each child to bring one hyacinth to school, or get them from the florist. They are usually available cut or potted in flower shops in early spring. You want nice, tall, full blooming ones. The red and white varieties seem to hold up better than the purple which get slimy and soft a bit more quickly, but all kinds will work. If you tell the florist that you are a teacher and that you want to buy a large quantity, perhaps he or she will give you a discount. (Hyacinths are also wonderful bulbs to plant in the garden in the fall, along with crocus, narcissus and daffodils, to be enjoyed in the spring.)

2. Double thread your needles and tie a knot about four inches in from the end. This will give you room to tie the two ends of the necklace together when you are finished stringing.

3. Gently remove the blossoms one by one from the stems. If the children are helping with this, make sure they grasp each blossom at its base and gently wiggle or twist it off. You don't want them to be torn or bruised. Do this right before the stringing.

4. Thread the blossom onto the string, by pushing the needle up through the bottom center of each blossom. Slide the blossom gently down to the knotted end. Continue threading until you have about four inches left on the thread. Then tie another knot to finish. Make single color necklaces or experiment with multi-colored ones.

5. Tie the two ends of thread together to make the necklace. It's nice if it is a continuous circlet of flowers, but it's okay if there is empty space in the back or on the sides. Put a piece of tape on each necklace and mark it with the child's name. Keep the necklaces refrigerated if you are not using them right away. They will last a lovely day or two, depending on the type of flower used, but don't expect much more than that.

NOTE: If you make the thread for each necklace long enough so that it will slip over the child's head easily, it won't have to be untied. Slip it on over child's head as you thread her or his needle to assure a proper fit. If the children are making them for mom or another adult, make them a bit bigger.

Making Butter

Children love to eat butter. It's a wonderful moment when they experience that their efforts have transformed the cream into butter. A trip to a dairy farm to get the cream would really complete the picture.

You will need

- heavy whipping cream—one that is not ultra-pasteurized works much more easily. Try a natural food store.

- two or three clean marbles

- a clean, clear container with a tight fitting lid

- spatula

- plate

- wooden spoon

- optional—clover or parsley

What to do

1. Pour the cream into the container and add the clean marbles. Put the lid on the container, and be sure it is closed tightly.

2. Begin to shake the contents up and down and side to side. Keep shaking until the cream turns into butter. It will get lumpy and the whey (a watery, milky liquid) will separate out. The time required for the butter to come varies depending on the temperature, the kind of cream used and other variables. Just keep at it and be patient. The children enjoy shaking the container and listening to the sound of the rattling marbles get more and more muffled as the cream thickens. Chant or sing (make up a very simple tune) the following verse while doing the shaking:

   > *Come, butter, come.*
   > *Come, butter, come.*
   > *(Jenny's) at the garden gate,*
   > *Waiting with a butter plate.*
   > *Come, butter, come.*
   > *Come, butter, come.*

   (Substitute the name of the child who is shaking.)

   This rhyme can also be used as a way to gauge the length of the shaker's turn, particularly if you have a lot of children waiting to take a turn. Go through the verse twice for each turn.

3. When you have butter, scrape it out of the container with a spatula and put in on a plate. Then "paddle" the butter; pat it all around with the flat back of a wooden spoon or other utensil to shape it into a little butter cake and to paddle out any whey.

4. Decorate the little butter cake with tiny blossoms or green leaves, for example, clover or parsley. Some people press "butter stamps" onto the cake which leave decorative impressions.

5. Use the butter with bread—preferably some you have baked—or crackers for snack. Homemade butter is a special treat for a festival meal.

NOTE: During free play, we set up a little butter churning station with two chairs at the end of a table and the materials. The children could come and help as they wished. I would sit in one chair and the "shaker" was in the other. A clean glass jar (labels removed) is easy to see through, and the top won't come off with all the shaking. Choose a jar that is small enough to fit into the children's hands comfortably, yet large enough to let the marbles really rattle around and agitate the cream. If you use glass, be extra careful and wipe the outside of the jar periodically so it doesn't get slippery. Small wooden butter churns are available from Basketville . (See the Resources section.)

Washing Wool

A field trip to the farm to watch a sheep being sheared is a wonderful spring experience. Have a picnic lunch there, and bring some of the newly sheared wool back to school. Then begin happy days working with wool. If you take the children to a farm to see a sheep sheared, be sure to speak to the farmer first about how it will be done and how the sheep feel about it, so you will be ready to reassure the children. The job is often done with electric shearers, and sometimes the sheep are nicked and bleed a little. They aren't always so cooperative at "haircut" time. A wonderful story to tell or read to the children when you begin working with wool is Pelle's New Suit by Elsa Beskow. It tells the entire story of a new wool suit, starting with the lamb.

You will need

- raw fleece

- wash tubs

- drying rack or clothesline

- basket

What to do

1. If you can make a trip to a farm to get some fleece, that would be great. Otherwise, check with your local 4-H or extension services about nearby farmers raising sheep, or try weaving or yarn shops. Just be sure to get "raw" fleece that has not yet been cleaned or carded. Wool batting or stuffing has already passed this point.

2. Set up two wash tubs, one with a very mild soap such as one for washing wool—not detergent—and one with clear rinse water.

3. Pick through the fleece to remove large bits of field debris, sticks, burrs, etc. The children are often amazed at how many things stick to the sheep's coat.

4. Tear off handfuls of the fleece, squish them in the soapy water, then rinse them in the rinse water. This is fun to do outdoors during outside play time. If you do the washing outside, then you don't have to worry about wet floors and spills. It's also interesting to note how oily the wool feels. This is the natural lanolin in the wool—it gives the wool its characteristic "sheepy" smell. Don't try to wash all the lanolin out of the wool. Leave some in the wool, as it makes it easier to work with.

5. Spread the little clumps of wool to dry on the clothesline or drying rack.

6. When the wool fleece feels thoroughly dry, which will depend on the size of the clumps, whether you spread them out a bit to dry and the weather conditions, gather it into a large basket.

7. It's now ready for the next step, carding.

Carding

The purpose of carding wool is to get all the fibers of the wool going in the same direction so it can more easily be spun into yarn. It's a pleasant activity to do—indoors or out—and the children love the little fluffy clouds of wool that result.

You will need

- wool carders
- basket of washed, raw wool
- second basket

What to do

1. It's nice to have a pair of real carders for the teacher to use, but they are often a bit large for the children to use comfortably. I have had great success using small wire-toothed brushes which pet shops sell for brushing animals. They are not too expensive, so you may want to buy several pairs, as lots of children will want to help. The brushes will last for years.

2. Set up a carding circle—a chair for you and several small chairs and sets of carders for the children—around a basket of washed wool. A second basket holds the carded wool when it's ready.

3. Take a small bit of wool—you will gradually be able to gauge how much to take—and stretch it over the outside edge of one carder. Holding the carder with the wool brush side up on their laps, the children "brush" that wool with their other carder. They should always move in the same direction, not go back and forth. Brush and lift, back to the beginning, brush and lift... The carder holding the wool does not move very much, the other one does. The children

brush each bit of wool until there are no clumps and it becomes a fluffy little cloud of wool. You will see that if you put too much wool on the teeth of the carders, you will not be able to brush through it. If you put too little on, it almost disappears as you card it.

4. Place the carded wool into the second basket. This is an activity that can go on for weeks. In fact, I often keep wool for carding available as an activity choice during free play time. It's just the thing for children who need some sitting down or who can't figure out what to do.

5. You can also card the wool—though not as thoroughly—by teasing and spreading out the little lumps with your fingers. This will sometimes satisfy little fingers until they have a turn with the carders.

NOTE: It is not necessary to have a lecture-demonstration lesson about carding wool with little children. They may come and go while the carding is being done, but they truly enjoy just having the work go on in their presence. It gives them such a sense of satisfaction, as if they were doing it themselves, and they will learn all they need to know just by being there, by carding the wool themselves and by seeing firsthand the process by which a sheep's wool begins to become something to wear.

Spinning Wool

You can actually spin wool with your own hands! But if you are lucky enough to know how to use a spinning wheel or drop spindle or can invite a spinner into your classroom, the children will benefit by seeing the wool actually become yarn. Drop spindles are not so hard to use—I taught myself by following the directions from a book and practicing a bit before trying it in front of the children. It's fun!

You will need

- carded wool

- drop spindle—check a weaver's shop or a yarn shop

- spinning wheel—get someone to come and spin in the classroom. The children love to see a spinner at work.

What to do

1. You can actually "spin" wool into yarn by rubbing little carded handfuls of it back and forth in the palms of your hands. You might not be able to knit a sweater with the results, but the children will get the idea.

2. Drop spindles are not so difficult to use. You could teach yourself to use one. This is not something for the children to do, rather they can watch it being done, and know that people invented tools, such as the drop spindle and the spinning wheel, to help them work more effectively. They needed lots of yarn to make their clothes, blankets, etc.

3. Contact a weaver's shop or sometimes a yarn or craft shop to learn about someone locally who spins yarn. Invite a spinner into your classroom, and the children will not only see the carding being done but also what comes next. Again, with little children, just experiencing the activity is enough. It's as if they were seeing straw spun into gold! Have the spinner set up and work during free play time, and all the children will gather round. Pure magic!

Weaving with Wool Yarn

Weaving is a more complicated activity and as such is appropriate for the oldest children, but it is also wonderful for the younger children just to watch and experience the process.

You will need

- a branch with a V-shaped crotch that is not too wide—for a loom. Gather these over time, perhaps during walks in the woods

- heavy thread or string for the warp

- scissors

- wool yarn—from a yarn store or, preferably, yarn that you have spun

- optional—feathers, stems of grass or weeds, brightly colored wool

What to do

1. String the warp thread onto the branch loom by tying it to the left side of the branch near the bottom of the "V." (The warp are the threads that form the basis for the weaving.)

2. Bring the thread (or string) over to the right branch, wrap it tightly all the way around the branch several times, then go back and around the left branch, leaving about a 1/2-3/4 inch space between the strings. Continue doing this back and forth wrapping until the crotch of the "V" is completely strung, leaving an inch or two at the top end of each branch unstrung. This "margin" will keep the weaving from slipping off the ends of the branch.

3. Begin weaving with the yarn at the left-hand side of the "V", going alternately over and under the warp threads. You may either cut the yarn to the length you need to weave one strand from the bottom to the top of the warp, or you may use whatever length yarn you have and weave continuously, wrapping around the topmost warp thread and weaving back down, always alternating over and under the warp threads. When you come to the end of your yarn, tuck the end in, take a new strand, overlap it a bit over the strand you just finished weaving, and continue from where you left off. Try alternating or changing colors. If you weave with the individual strands of yarn, tuck the ends in for a more finished look.

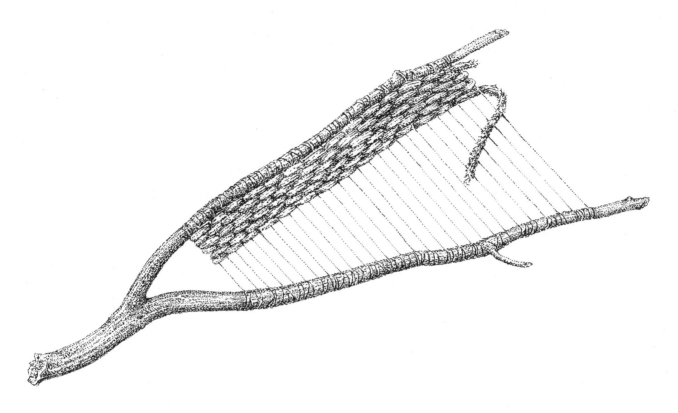

NOTE: A more simple weaving can be done by using carded unspun wool. Just take a handful of carded wool and roll and form it loosely into a "log" shape. Weave with this, adding new wool as necessary. Using this "fatter," unspun wool is often easier for little fingers than yarn.

4. As you are weaving, be sure to press the vertical strands of yarn close together. The more closely you press them the tighter your weave will be.

5. Weave until you reach the right-hand side of the "V" and the loom is full. Weave in a feather, a long stem of grass or a bit of brightly colored wool as a decoration. Use your imagination.

6. If you have woven fairly tightly, you can cut the weaving off the loom by carefully snipping the warp threads and tying the warp ends together. (This is why you wrapped the warp ends several times around—to leave enough for knotting.) The weaving will also look beautiful as a wall hanging just as it is on the branch, especially if you add some decorative elements as mentioned above.

Picking the Spot

Working with the Earth to create a garden, however small, is an experience that every child should have. To be able to tend and care for the Earth in this way engenders attitudes of stewardship. The rewards of this caring go far beyond the harvest of vegetables, flowers, fruit and fun. It really is true that children are much more likely to eat things they've picked from their own garden. And this realization that the food we eat comes from the Earth—as a result of the labors of humankind—nurtures their sense of security and the wholeness of life and allows them to feel gratitude for the gifts the Earth so generously gives. Encourage the development of these sensitivities by taking the time and making the effort to garden with your children. Gardening is probably the single most important thing that you can do to make children more environmentally aware.

You will need

- a sunny, outside area

What to do

1. Choose the sunniest spot that is available to you as the location for your garden, whether it is a corner of the playground, a small area under your classroom windows or a vacant lot down the block. If you are working in an urban location with a concrete or asphalt play area, choose a sunny spot to become a container garden.

2. It's nice if the garden is located adjacent to or in the area where your children play outdoors as this allows the garden work to take place during outdoor play time.

3. It's also nice if the area is a somewhat protected area, to keep it out of harm's way as much as possible. It is always a possibility that mischief-makers—both animal and human—may try to undo your work or help themselves to your produce. Here's where an attitude of perseverance, a bit of forethought and a sense of humor are called for. The way you respond to disturbances of this sort—with a sadness for what's been harmed but an unyielding determination to keep at it and make it right—will teach your children more than you can imagine.

4. Once you've chosen the sunniest available spot, don't neglect to check with your local "powers that be" to okay your gardening idea.

Preparing the Soil/Constructing the Garden

When constructing the garden, remember that you can start small and get bigger as you go along. Your garden needs to have beds in which you plant your seeds and paths between the beds. You can make your beds just wide enough so that a child can reach to the middle from a path on either side. That way they don't need to walk right where the plants are growing. For container gardening, you can use almost anything from large half-barrels to flower pots to window boxes to boards nailed together to make a frame. The important thing is that you have sunlight, enrich your soil and don't overcrowd your plants.

You will need

- garden area or large half-barrels, flower pots, window boxes and/or wooden frames, etc.

- shovels and rakes, ideally both adult- and child-sized

- well-composted manure, leaf mulch, compost, top soil, as needed

- buckets or wheelbarrow

- hand trowels

- optional—sand

What to do

1. Once you've chosen your garden area, you need to prepare your soil.

2. If you are using an area that is grassy at present, you need to take time to remove the grass using a spade or shovel. Dig in and turn up a shovel full of soil. Shake the soil off the clumps of grass and back into the garden. The remaining grass can either be composted or used to fill in bare spots on the lawn or play area. Continue to remove all grass from your garden area. Start small and clear additional space as you need it.

NOTE: If you plan a season or two ahead, you can thoroughly mulch your garden area with wood chips, leaf mulch or well-composted manure the fall before. The mulch will kill the grass and turn it into soil nutrients which can be spaded into the soil come spring.

3. In any case, you want to work your garden soil a bit to loosen and aerate it and to add nutrients in the form of well-composted manure, leaf mulch or compost. Eventually you'll be able to use your own compost! Also, if your soil has lots of clay which makes it very compact, you can bring a few buckets full or a wheelbarrow load of sand over from the sand pile and work it in. All this is aimed at giving you rich, dark soil which is crumbly and easy to work. You don't need to be too scientific about how much of what you add—just don't go overboard on anything and work your additions in well with shovels and rakes.

4. If you will be using containers for your garden, get large piles of materials such as the manure, leaf mulch and top soil dumped onto the playground in the area where you want to garden. Then gradually begin mixing them together to make your working pile of soil.

NOTE: Composted manure is available just about everywhere. My husband was able to get it in New York City for a garden project he worked on when teaching there. Try the police stables or parks. The parks department may also be a good source for leaf mulch. Tell them what you are doing with it.

5. You can be creative with your garden space as well. We once made a lovely, round garden that the children loved to walk in.

Choosing What to Plant

If you are a teacher who does not work with your children through the summer, you will need to choose vegetable varieties that will sprout and mature quickly so that you can enjoy them with the children before school is out in June. You may also want to choose varieties that will be ready when you return in the fall, for example, pumpkins and popcorn. If you are teaching throughout the summer, you have the advantage of lots of time and, also, a built-in solution to the problem of garden care over the summer months. in fact, the garden can really become a focus of your summer program.

You will need

- vegetable seeds such as radish, lettuce, carrots, peas, pumpkins, popcorn, beans

- flower seeds such as zinnia, marigold

- herb seeds such as mint, camomile, parsley, lavender, sage, thyme

- hand trowels

- watering can

- optional—eggshell halves

What to do

1. When looking for vegetable varieties that will mature quickly consider:

Radish—the fastest growing vegetable—often ready to eat in three weeks or so. There are the round red and long white varieties. They are easy to grow and wonderful to harvest.

Follow package directions for sowing. Some children love to eat them, others find their spiciness a bit much. Try cutting them into slivers.

Lettuce—try the leafy varieties as these mature more quickly than head lettuces. There are many different kinds and they do well when planted with radishes.

Peas—these can be started early on as they like cool weather. They come in climbing and bush varieties so plan accordingly for whichever you choose.

2. If you will be around during the summer, you can choose to plant just about anything you like. If you will be gone from your garden until late summer or early fall, plant things that you can harvest then, and pray for rain. We used to plant flowers such as zinnias and marigolds, herbs such as mint, camomile and parsley, and vegetables such as pumpkins, winter squash and popcorn. We mulched our garden heavily before our summer leave-taking to keep down weeds and keep in moisture. In the fall when school resumed, we weeded the garden and got it back

into shape. The flowers kept producing blooms that graced our Season's Garden, and we waited until the first frost to harvest our popcorn and until close to Halloween to bring in our pumpkins. One year our biggest pumpkin (and it was gigantic!) was a "volunteer" which sprouted in our compost pile.

3. When choosing what to plant, keep in mind the idea of companion planting. This means growing next to each other certain plant varieties that do well together by nourishing each other and/or by discouraging harmful insects. Marigolds are well known insect inhibitors and make a fragrant, colorful garden border. They also enliven your autumn Season's Garden and make a sunny yellow natural dye (simmer the flower heads) for cloth or wool! Other plants that do well together are:

- carrots with peas and lettuce

- tomatoes with parsley

- parsnips or potatoes with peas and beans

- corn with pumpkins

4. Once you've decided what to plant, follow the care and planting instructions on the seed packets, and seek additional guidance and advice from gardening books and experienced gardeners (maybe parents).

5. Remember that many seeds can be started on a sunny window sill indoors several weeks before you plant

them outside. This works especially well for tiny seeds such as herbs and lettuce. Plant them in saved eggshell halves. When you're ready to transplant, just gently squeeze the shell to crush it and set the whole thing, plant, shell and all, into the garden. The shell actually provides some nutrients, too.

6. All vegetables enjoy the presence of aromatic herbs such as lavender, sage and thyme which are a delightful addition to your garden. The herbs can be used to brew snack time teas and hung in bunches to dry for cooking and making sachets.

Caring for the Garden/Tools You'll Need

Once you've gotten the garden going, it can provide the children with a focus for their work outdoors. Weeding and watering are daily tasks. Mulching should be done as needed to help control the weeds and to help keep moisture in the garden. Your garden should be a haven for helpful insects such as ladybugs, praying mantis and earthworms. Encourage the children to appreciate and protect these creature by handling them gently yourself. These creatures should always be welcomed.

You will need

- shovels and rakes, adult- and child-sized

- hand trowels

- watering can(s)

What to do

1. When choosing tools, especially child-sized ones, go for quality over quantity. A rake and a pointed shovel are the most basic. Hand trowels are helpful, too. Many hardware and garden stores carry child-sized tools—just be sure that what you are purchasing is sturdy and capable of real work. A word of explanation to shopkeepers or a plea to parents and the school community might bring donations of tools for your gardening program. It's better to have a few good tools and take turns than lots of plastic ones that will soon get broken.

2. You'll also need a good watering can—and it helps to have more than one of these. If you don't have an outside faucet nearby, consider catching rainwater in a barrel placed under a downspout. It is also very important to establish good attitudes and habits of properly caring for your tools. Clean them and bring them indoors or put them in a tool shed after each use.

3. The most important thing to remember about gardening with children is that you **can** do it—it's not so hard. Just go ahead and do it; you'll learn together as you go along. Planting seeds in your children (as you do in your garden) of nurturing and caring for the Earth, companion planting, helpful insects, all these things which will impress their feelings and habits at a young age, can mature into clear, conscious ideas and attitudes when they are older.

Chapter 4
Summer

The Whole Earth Home and Classroom

Creating A More Natural Outdoor Play Space

The outdoor play area is a space that can easily be taken for granted, but it offers many opportunities for enhancing an awareness of nature and for creating interesting spaces for work and free play. As was true with using natural materials for toys and playthings indoors, many of the materials that can transform the outdoor area are free for the asking and will work just as well in an asphalt school yard and in your back or front yard as they will in an open field. It just requires a little time and effort to track them down and get them where you want them.

Stumps and Logs

As with indoors, stumps and logs—even larger than we could use indoors—make excellent additions to the play area. Logs, whole or split lengthwise and placed in interesting arrangements, are wonderful for climbing and walking on, and encourage children's creative play in many ways. We had a wonderful old log lying on our playground. It had a twisted branch at one end that was sometimes a dragon, other times a boat, a train or a tree house. A log, especially a split one, placed over a depression in the earth (dig one with the children if there are none) makes a wonderful bridge. You could line the "river" under it with stones.

Shorter logs, two to three feet in length, can be used to transform the sand area. Children love to play in sand and little sandboxes are often too small to really **play** in. It's nice to be able to accommodate the vigorous play of lots of children with a large sand area. Lay the logs end to end in a large circular form. Introduce a standing stump every now and then that can be used as a table or "oven." The sand is then dumped into the center and fills the form. Some of it will squeeze out between the logs, but most of it stays put. The logs are good seats, and the whole construction is great for walking and balancing on. A sand pile this size allows for all kinds of play, from real digging with child-sized garden shovels to small building projects and cake bakeries.

Smaller logs or branches (approximately four to six inches in diameter) can be used to build a nice climbing structure. The thicker pieces are sunk into the ground upright to provide good support. Then shorter and longer pieces are bolted onto the uprights at various heights and angles to connect them into a climbing structure. Very interesting to look at, and fun to play on!

Wood Chips

This natural material can transform an asphalt play yard into a forest underfoot! Often wood chips are available free from utility companies that trim around wires and poles. We have often gotten them by just speaking to the operators of the chipping equipment when we saw them working in our area. They were happy to have a place to dump a load or two. We literally carpeted over an asphalt playground with wood chips that are now two to three feet thick. We marked out paths with thin logs and filled in everywhere else. Once we built up the base, we had several wood chip "mountains" which we left intact and which provided lots of possibilities for running and jumping and a change of view from our otherwise flat play area.

The Garden and the Compost Pile

The garden was an important part of our outdoor play area, and the compost pile stood in a quiet corner. I've discussed setting up these areas in other parts of the book, but it is important to mention what an integral part they played in our outdoor time each day.

For many teachers, outdoor time is an unstructured time, a time to let go, relax, maybe talk with colleagues. Although these are important, outdoor time is also a good time for work for both teachers and children. After a chance to chat a bit, we'd take the brooms and sweep the walks, or shovels or rakes and go into the garden to plant or weed. Child-sized wheelbarrows carried the weeds to the compost pile; watering cans were filled and used to give our plants and flowers daily drinks. We also did our harvesting at this time. Since the garden was so close to the sand pile and play area, supervision of the children was not a problem.

The other thing that was interesting to note was that as the teachers worked earnestly, so did the children. Some of them would immediately come to do whatever the teacher was doing. While having good quality, child-sized equipment available facilitated their work, it became clear that even those who weren't helping us directly were much more focused and earnest in whatever they were doing when their teachers were involved in real work. Children learn an immense amount from doing what we do. They also learn more than we can imagine from **how** we do what we do.

Landscaping

Once you get your basic outdoor play space organized and naturalized, you can begin to add the landscaping touches that will make it really special. There is growing interest in landscaping with plants that provide food of various kinds from fruit to nuts, and in plantings that will attract birds and other wildlife. These both provide possibilities for bringing the children into closer contact with the wonders of the natural world and, ultimately, developing an appreciation of and for it. There is a wonderful mail order edible landscaping company in Virginia that is happy to help with recommendations and advice and to send you plants: Edible Landscaping (address in Resources section).

There are also books and organizations that can help in this regard. A good book is *How to Attract Birds to Your Garden* by Dr. Noble Proctor (see Resources section for details). Organizations that can help are The National Audubon Society, The National Wildlife Federation, and The National Arbor Day Foundation (addresses in the Resources section).

Many of these plantings can be done in raised beds or boxes and in containers of various kinds, so don't let an urban location stop you from "greening" your outdoor play area. It **can** be done, and the benefits will surprise and delight you. This is also a good area in which to solicit parents' help and support. Bake sales to get some funds to buy trees and plants and potluck work days to get the transformation underway can go a long way toward building a sense of community and fostering an understanding of what you are trying to bring to the children.

Bringing Nature In:
The Season's Garden in Summer

For more detailed information about setting up a Season's Garden, see the section on "Creating a Seasonal Garden" in the fall chapter.

Colors: Warm, fiery reds, oranges and yellows.

Special Objects: Shells, sea objects, flowers or weeds that are going to seed, summer produce of all kinds—melons, a bowl or basket of cherries or berries, cherry tomatoes, etc.

NOTE: Food placed in the Season's Garden should not remain there indefinitely. It can be placed there one day and brought to the children's attention, perhaps with a short story about where it came from or how it grew. Enjoy it for snack or use it in a cooking project before it has a chance to spoil.

Wreath: Your wreath can become a flower wreath in the summertime. Using a base of straw, wild grasses or weeds, you can bind on all kinds of flowers, wild or cultivated. Many flowers actually continue to look quite beautiful even after they've dried and faded, so you don't have to constantly replace them. It's also nice to add some little bumblebees, as these are familiar summer visitors. Shape them from yellow wool or beeswax, adding a bit of dark color for contrast (even dark blue or purple will do as black is not always available) and, using string, hang them from the wreath.

Plantings: It's fun to turn your dish garden into a tiny beach for summer, and it gives you a chance to remove the soil, moss, grass or whatever else you've cultivated during the year so that you can start fresh in the fall. The moss, bulbs or plants can be replanted outdoors in appropriate spots or, if necessary, put in the compost pile. Add the soil to the compost as well. Fill your container about 1/2-3/4 full with sand. You can add a shell or small saucer of water, if you like. Then place different kinds of sea shells, special stones and other sea objects in the sand. Pieces of beach glass (smooth, weathered pits of broken glass you often find washed up on the beach) are fascinating to the touch. The children will love to come and sift the sand through their fingers and play with the shells.

Dandelion Chains

This can be done indoors, but really fits during outside play time. You can substitute other flowers, though dandelions are readily available—they even grow through cracks in the sidewalk! Besides being beautiful crowns, the chains make lovely decorations for your Season's Garden or castle decorations in the sandpile. The children love to watch you make something from "nothing."

You will need

- dandelions or other flowers with tubular stems

What to do

1. Gather dandelions. Children love to do this—just show them how to pick them so you get a nice long stem.

2. Use your fingernail to make a small slit (1/4-1/2 inch) in the dandelion stem about one inch below the head of the flower.

3. Slip the next dandelion through the slit, stem end first, and pull it through until it stops at the flower head.

4. Continue making slits and attaching dandelions until the chain is the length you want. To make a crown, attach the last stem to the beginning of the chain. You can twine it around the first flower or knot it to the stems.

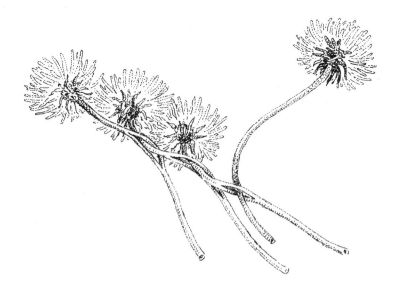

Shooting Star Streamer Balls

This sewing project is a bit difficult, but the results are a wonderful summer toy. You may want to just make a few for your class to play with outdoors. If you are very industrious and/or have access to a sewing machine, they make great leaving-for-summer gifts for the children.

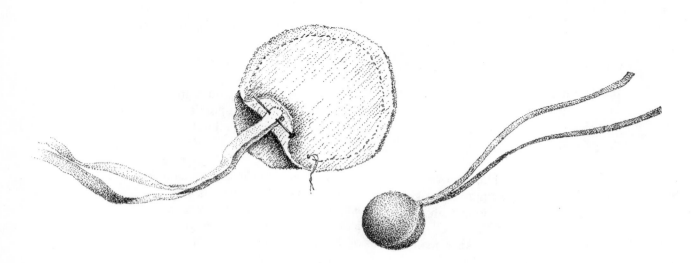

You will need

- sturdy fabric—solid colors are best, red cotton knit or red wool felt are especially nice

- scissors

- needles and thread

- rice or sand for filling

- ribbons—two to three per ball— each approximately 18 inches long in hot colors: red, orange, yellow, gold

What to do

1. Cut the fabric into three inch circles. You will need two circles for each ball.

2. On a machine or by hand using a small, sturdy stitch such as a backstitch, sew the circles together about 1/4 inch in from the edges, leaving an opening for filling. The older children will be able to sew this by themselves, but younger children or those less experienced in sewing will need assistance.

3. Overlap the streamers and pin them to the inside of the seam allowance of the opening. Stitch them firmly in place (see illustration).

4. Turn the entire ball "inside out" (you have been working on the inside) so that all the stitching is now on the inside and the outside is out!

5. Fill with rice or sand (not too full), fold in the edges and stitch up the opening. It is important to stitch this carefully so the shooting star doesn't leak!

6. Let the children take them outside and throw them up into the air or back and forth to each other. The ribbons will stream behind the shooting star as it flies.

NOTE: While some children may not be able to do the stitching required for these, as it needs to be small and close together so the filling won't leak out, they love to choose their ribbon colors and do the filling. A nice idea is to make them at an end of the year (or the beginning of summer) festival. Invite the parents, then each parent can make one with his or her child. It still helps to have the circles sewn together and the ribbons pre-cut. Then let them take it from there.

Butterfly Crowns

These unusual crowns also make beautiful window pictures. The children can wear their crowns for an outdoor summer festival—a part of which might be acting out the poem included below.

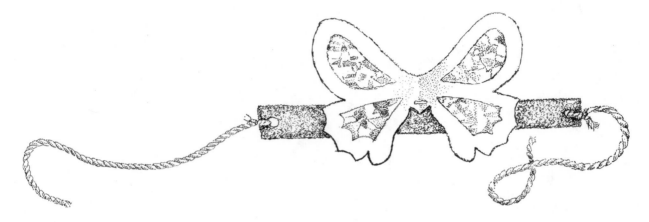

You will need

- construction paper—any color except black or brown

- scissors

- clear contact paper, or tracing paper or waxed paper and glue

- hole punch

- yarn or ribbon for crown ties—cut in pieces approximately 6 inches long —two per crown

- scraps of tissue paper—light colors

- stapler

What to do

1. Fold an 8 1/2 x 11-inch piece of construction paper in half from top to bottom and cut the butterfly shape on the fold.

2. Cut out the center of the wings on both sides (see dotted lines) to make the frame.

3. Cut clear contact paper to cover the hole in the wing frame and attach it to the back. Now you have a butterfly with a sticky space for a wing!

4. Prepare the crown headbands by folding an 8 1/2 x 11-inch piece of construction paper two or three times lengthwise to make a one inch band. Crease the folds well. Use a hole punch to make holes about one inch in from each end of the band. Tie a piece of yarn or ribbon through the holes.

5. Give the children the butterfly shapes and scraps of colored tissue paper. They tear off small pieces of the tissue paper and stick them to the contact paper. Encourage them to fill

the wing space completely; overlapping tissue paper pieces is okay.

NOTE: You can also pre-cut or tear the tissue paper if you like. Tearing is nice as it gives the tissue paper a soft edge. Just be sure that the children tear off small pieces of tissue paper so they don't fill the entire space with one color.

6. Staple the center bottom of the finished butterfly to the middle of the headband. Make sure to match the color of the headband to the butterfly frame, so a yellow headband goes with a yellow butterfly, etc. Mixing the colors of the frames is distracting and makes the finished project less beautiful.

7. The following poem holds lots of possibilities for play.

A tired caterpillar went to sleep one day,
In a snug little house of silk and grey.

He said as he softly crawled into his nest,
Ah, crawling is fine, but rest is best.

He slept through the winter long and cold
All tightly up in a blanket rolled.

'Til at last he awoke on a warm, spring day
To find that winter had gone away.

He awoke to find he had golden wings
And no longer need crawl over sticks and things.

Ah, crawling is fine said the glad butterfly
But the sky is best when you learn to fly.

Unknown

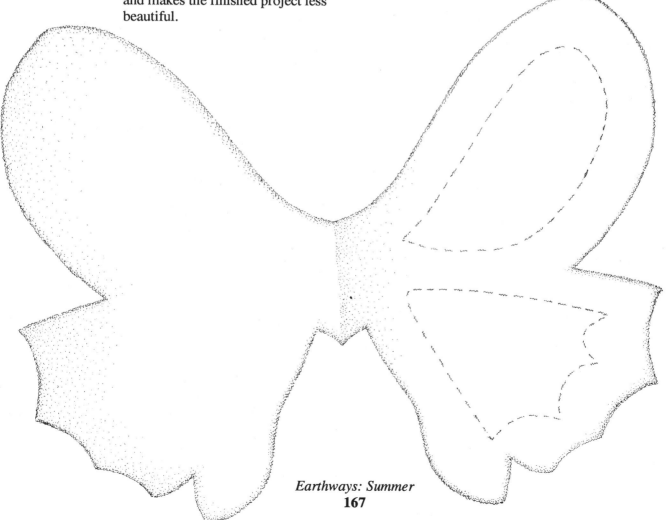

Walnut Boats

These are very easy to make and can be used indoors, outdoors in a tub or water table or in the creek if you are lucky enough to have one nearby!

You will need

- walnut shells—you want perfect half shells, so crack them carefully

- small bits of beeswax or clay (beeswax works better as it doesn't dry out)

- string cut in 9 to 12-inch lengths

- paper scraps

- scissors

- toothpicks

- water table, tub, large dishpan, or creek!

- optional—buttons

What to do

1. Crack the nuts and have the children help you remove the nut meats. This may mean snack time to them! Save the rest for a baking project.

NOTE: A friend told me that an oyster shucker is great for opening walnuts when you don't want to break the shells. Just stick it in the little opening often found at the flat end of the walnut and wriggle it open.

2. Place one end of the string in the bottom of an empty walnut shell and press a small ball of beeswax (or clay) over it to hold it in place.

3. Cut a tiny sail shape from a scrap of paper and poke a toothpick through it. Now the boat has a sail on its mast.

4. Stick the bottom end of the toothpick into the ball of beeswax.

5. If you want, attach a button to the loose end of the string. This gives the children more to hold on to as they pull their sailboats around the water.

Special Water Play—Creating Mossy Islands

Creating this play space is fun and easy, and once created, it becomes a little world in which children can spend countless happy hours of play. If you can bring it outdoors, that's even better.

You will need

- moss—collect a bit from where it grows naturally: moist areas in the woods, stream banks, etc.

- small shovel or garden trowel

- large container to hold water—sand or water table, galvanized tub, basins

- large and medium-sized rocks

- sand

- water

- walnut boats from the previous activity or corks, toothpicks and scrap paper

- plant mister

- optional—beeswax, sticks and string

What to do

1. Remove the moss carefully with a layer of dirt and keep it moist in transport. Repair the area from which you removed the moss by filling holes and patting down loose soil. Don't take too much from any one place.

2. Arrange the rocks in the center and corners of the water container, building them up to make little cliffs or

mountains. (Create caves by leaving spaces between the rocks.) Make a little beach near the bottom of the cliff by putting some sand on top of a flat rock. If you are using a small basin, you may want only one "island" in the middle.

3. Carefully add the moss to the rocky islands, making them green and lively. Make sure the moss has a good foundation of soil, adding some if necessary. Pat and fit the moss into the rocky spaces. You don't need to cover the rocks entirely. These are the mossy islands, and depending on the amount of moss you have been able to collect, you may want to have only one mossy island and leave the others rocky.

4. Fill the container about half full with water. You want enough water to make it fun to play in without flooding the islands and beaches.

5. Allow the children to play in the water with the walnut boats. You can add little beeswax people who live in the rocky caves and visit the beaches. Corks (usually available in bags of assorted sizes at the hardware store) with toothpick and paper sails also make nice boats, buoys or channel markers. The children can gather small sticks that can be lashed together to make docks or rafts.

6. Keep the moss moist by misting it daily. When disassembling the islands, the moss can go into your Season's Garden, or patch damp areas of your playground or return to its natural home.

Bark Boats

A very simple little boat with a leaf sail—the kind our grandfathers may have sailed down streams years ago.

You will need

- chunks of bark in varying sizes from 3-4 inches long to as big as you can handle. Pine bark (mulch) is available at garden shops.

- sticks

- leaves

- an oyster shucker, awl or ice pick—some kind of tool with which to bore a hole

- optional—small cup hooks and string

What to do

1. Bore a small hole in the center of the piece of bark to hold the mast for the sail. If you have a very large piece of bark, you may want to have more than one mast. Have the children bore the holes. Depending on the sharpness of the tools they're using, they will need more or less supervision.

2. Thread a large leaf onto a stick, and place the stick in the hole.

3. If you want to attach strings so the children can pull their boats (or not lose them downstream!), screw a small cup hook into the front edge and tie on a string.

4. Sail the boats in a big tub of water or water table or, better yet, take the children to a creek or stream and really let them sail!

Parachute People

A fun and easy way to explore the air. Take these to the top of the climbing structure and see what happens.

You will need

- 8 —10-inch squares of lightweight cloth or tissue paper, white or colored. Cheesecloth will also work as long as the weave isn't too loose and open.

- 12-inch lengths of thread, four per parachute

- beeswax or clay

What to do

1. Tie the thread to the four corners of the cloth by scrunching up each corner, wrapping one end of the thread around it a few times and tying a knot. Leave the other end of the string free.

2. Attach the other three strings. Gather the loose ends together and tie them into a knot.

3. Embed the knot in a ball of beeswax or clay (a pencil point helps with this), and if you like, form a little person. Just make a very general shape; don't make it too detailed.

4. The older children will be able to make these on their own. The younger ones will need some help or may just want to play with them after you've made them.

NOTE: You can help the children cut their own string by taping a piece that is the correct length on the table. They can use it as a measuring gauge.

Paper Birds

The air is full of birds in the summertime. These little birds with folded paper wings can hang from a string or be attached to the end of a stick. The children love to "fly" them.

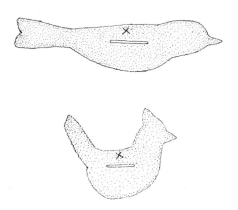
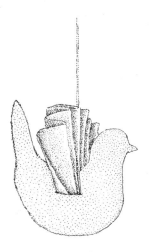

You will need

- colored construction paper
- scissors
- drawing paper or tissue paper— white or colors
- crayons
- string

What to do

1. Cut the construction paper into bird shapes. Use the large or small shapes illustrated or invent your own.

2. Cut a small slit in the body where the wings should be.

3. To make the wings, cut a small rectangle from the drawing or tissue paper. If using drawing paper, you may have the children color the "wings." Fold the rectangle length-wise accordion style (back and forth). It should be wide enough so that you can fold it approximately 10 times. Slide the folded piece through the slit in the body. It should be a tight fit.

4. Spread the folded wings a bit. You can also glue the top edges of the two wings together so that they stay open. Attach a piece of string to the body at the top of the back.

5. Let the children "fly" the birds by holding the string or tying it to the end of a stick. Just take care that the sticks are used carefully.

Moving Pictures

Children love pictures that move, as the movement makes things come alive. Your oldest children will enjoy making these themselves; the younger ones can help you make some for the class to use.

You will need

- drawing paper—heavy stock—at least 8 1/2 x 11 inches

- crayons

- scissors

- tape

- support sticks—popsicle or craft sticks, tongue depressors or strips of smooth cardboard

- optional—glue

What to do

1. Have the children draw and color a simple sea scene on the paper. Basically you want to include the sea, the sky, some fish, sea birds, etc. Don't just make line drawings, but fill in the area with color.

2. Cut a slightly curving line through the paper in the sea, making sure to leave a one inch margin on each side of the paper. Place a piece of tape at each end of this curving cut on the back side of the paper for reinforcement. Also be sure to cut it no higher than the length of your support sticks (keep the cut in the bottom one-third of the paper). The stick needs to reach the cut and still be longer than the bottom edge of the paper.

3. With the children's help, draw a small sailboat on a separate sheet of paper. Heavy weight paper, e.g., watercolor paper, works well for this. The boat should be a size that will fit in with the scene—not so big or so small as to seem out of place.

4. Cut out the boat and glue or tape it to one end of the support stick. The boat should be wide enough to cover the width of the support stick, so the support stick does not show.

5. Slip the stick through the hole and down past the bottom of the paper. Set the boat into the slot so that it appears to be part of the scene.

6. You can add extra support to the scene by gluing a piece of paper the same size to the top and two sides of the back of the scene. Leave the bottom open for inserting the stick.

7. Sail the boat along the sea, perhaps stopping to swim or to look for a whale! Endless adventures can develop from such a simple starting place.

8. As you can imagine, this simple technique can be adapted for many different kinds of scenes. You could have a little child on your stick, walking in the woods or up a mountain, or an animal wandering through a meadow. Experiment with new possibilities. The more difficult ones may be too hard for the children to do, but they will love to watch and help you make them for the classroom and to play with them when they're done.

The Bean Tepee

Children love to build their own houses, and summer is the perfect time to move this satisfying activity out-of-doors. If you don't have branches and sticks on your yard or playground, import them. Your efforts will be well rewarded. This house will not only provide a place to play, it will provide a snack as well!

You will need

- garden area or planting boxes

- shovel or spade

- sticks, branches or bamboo poles— at least 5 feet long

- rope or strong twine

- pole bean seeds—any kind will do although the scarlet runner bean variety has the added advantage of producing beautiful red flowers that attract hummingbirds!

- watering can

What to do

1. Prepare the area where you will plant the seeds. If you are planting them in the garden, plant them near the outside edge, so the children can have easy access to their house without trampling the lettuces! Also, as they are tall when grown, you may want to place them so they don't block sunlight from reaching other parts of the garden. You can also just dig up separate areas on the playground where you'd like the tepees to be. Using a shovel or spade, turn the earth in a circle large enough to accommodate two to three seated children (approximately 2-3 feet in diameter). Remove any grass, shake off the dirt and place the leftover grass in the compost pile. Smooth the soil, adding in some compost if you have it. If you are an urban gardener, you will need large low planting boxes or containers, about five of them, spaced in a circle with an opening on one side which will eventually become the door.

2. Place the stakes or branches in the ground around the edge of the circle, leaving an opening on one side for the entrance. You need to use at least three poles; four or five will allow for thicker foliage cover. Tie the tops of the branches together in the center—tepee-style. Bind them fairly well so they will hold.

3. Plant three to four bean seeds around the outside base of each pole and water well. Keep the earth around the seeds moist and they will sprout more quickly. The watering is an excellent activity for the children.

4. Once the seeds begin to sprout and grow, the plants will reach out with delicate tendrils and climb up the poles. This is an amazing thing to see. Be sure to share it with the children.

5. The tepee house will grow larger and more private as the plants grow. Use your judgment as to when it can be "moved into."

6. Don't forget to eat the beans and watch for hummingbirds.

The Stick House

This house is fun to make and can provide hours of industrious play.

You will need

- long branches or poles—at least five
- twine
- small branches, twigs
- grasses, vines, leaves

What to do

1. Help the children set up the long poles that will be the base of the house. Locate them by a wall, in a corner, against a fence or by a tree. Lean them up against whatever support you choose and wrap the tops together with twine. Tie them together tightly at the tops, or weave the twine in and out and around, allowing the poles to spread out a bit. This method works if the poles or

branches are supported by a fence. You can also, at this point, lash them to whatever support you are using—the crotch of a tree, top of a fence, etc.—to make the house extra sturdy.

2. Weave smaller branches, twigs, bits of vine, grasses—really anything—in and out of the long poles from the bottom up. Don't forget to leave a space for the door, and you can also leave holes for the windows. Just stop weaving when you come to the window space, wrap your weaving material around the last pole and weave it back the other way.

3. Weave in the smaller materials until you reach the top of the house. You can always go back and fill in the holes—mud works very well for this!

4. The children can do most of the house building. You will just need to get them started by helping them with the long poles, or perhaps only by providing the pile of materials and suggesting a house. Show them how to start the weaving, provide occasional help and consultation as necessary and let them do it!

SUPPLYING THE MISSING LINKS: OUTDOOR PLAYHOUSES **3+**

The Cloth House

We've probably all made houses of cloth, using old sheets, at some point in our lives. Have a basketful of cloths just for use outdoors and dip back into your childhood memories to get the children started.

You will need

- baskets of cloths of all sizes
- string, twine or rope
- clothespins

What to do

1. Take the baskets of cloths outdoors one day and begin to make a roof.

2. Help string twine where needed to provide places to hang the cloths. Bushes, trees, fences, etc., can all be foundations.

3. Provide technical assistance as necessary!

4. Unlike the Bean Tepee and Stick House, cloth houses are rebuilt each day. Have the children help you fold up the cloths when it's time to go inside. When the cloths seem to need a washing, bring water tubs, soap and scrub boards outdoors, string a clothesline and have a wash day!

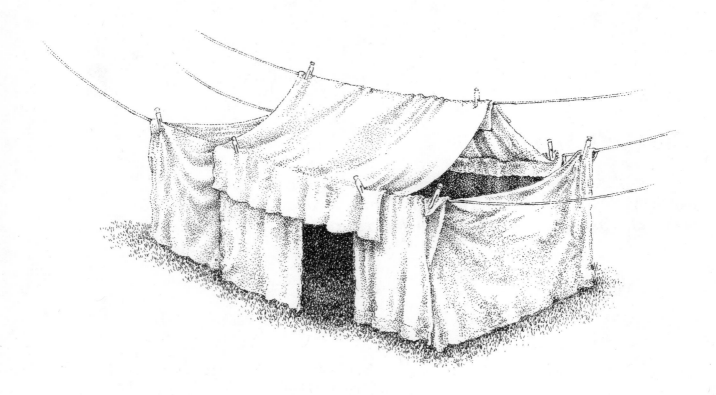

Berry Picking Plus

This is the time of the year to enjoy picking, eating and preserving all kinds of berries. The children often love to eat fruit and it is wonderful for them to experience where it comes from. If you made a trip to a pumpkin farm in the fall, check to see if it grows berries for picking. It's wonderful to go back to the same place and enjoy it in another season.

You will need

- a place to pick berries: raspberries, strawberries, blueberries, thornless blackberries

- containers

- masking tape and pen

- parents to accompany you

- signed permission slips

What to do

1. Make arrangements to visit a farm which has "pick-your-own" berries of some kind. Your pumpkin farm connection from the fall or local 4-H or extension office can probably help you locate one. When you call the farm for directions, be sure to ask if you need to bring your own containers. If you do, ask parents or your local grocery to help in collecting those green plastic or cardboard pint size berry containers. You will need one to two for each picker. Also ask the approximate cost for a pint of berries in case you need to collect some of the expenses from the parents.

2. Send a note home to the parents to inform them of your plan, to get their permission for their child to go and to collect any money.

3. Arrange to have enough adults accompany you so that each adult is responsible for no more than four or five children. If using private cars, each driver should have a map and directions, phone number of the farm, and a list of the names of the children in their group.

4. Prepare the children for their trip by telling them a story about how the trip will go. Be sure to include all the important details like how they will behave in the cars or bus, what kind of clothes they will wear (work clothes: sturdy shoes, long pants, sun hats and old shirts as berry stains don't come out easily), what they will do when they get there (stay with their group and, of course, never wander off), what the berries they pick will look like (e.g., strawberries will be deep red not orange or white), and how they will be careful not to eat too many or they might get a tummy ache!

5. When you arrive at the farm and receive your instructions for picking, be sure to show the children what a nice, ripe berry looks like so that they will pick good ones. Perhaps each child can pick two containers—one for home and one for the class, but don't count on it.

6. Give each child one container at a time to fill. Have a large shallow box (sometimes the farm will provide these, if not, bring your own) to hold the filled berry containers. Mark the containers with the child's name using masking tape and a pen, especially the one container they will be taking home. Marking the containers can be done before you leave school to save time and aggravation!

7. Pick lots—occasionally cautioning the children not to eat too many although you certainly should expect them to eat some.

8. Don't stack the berry containers on top of each other for the trip home. Otherwise you wind up with lots of squashed berries and lots of sad children.

9. On returning to school, refrigerate the berries that aren't going right home.

NOTE: Decide ahead of time what you will do in case of rain. If the farm will let you and the children wear rain gear including boots, go anyway! This can actually be fun; or cancel and choose another day.

Sample Note To Parents

Dear Parents,

On _____(*date*), our class will be taking a trip to_____
(*name of farm*), to pick _____(*kind of berries*). We will go by
_____ (*private cars or bus*) leaving school at _____(*departure
time*) and returning by _____(*return time*). The children will need to wear
appropriate clothes, including long pants (*saves knees and protects from bug bites
and scratches*) and sturdy shoes—no sandals, please. Berries do stain, so have your
child wear an old shirt. Please sign and send in the permission slip below along with
_____(*amount of money*) to cover the cost of the transportation and berries.

Many thanks,

_____
(*Signature of Teacher*)

My child,_____, has permission to accompany his/her
class to the _____(*name of the farm*) on
_____ (*date*). I understand that transportation will be provided by
_____ (*private car or bus*).

_____
(*Signature of Parent or Guardian*)

Berry Shortcake

Once you've picked berries, it's nice to use some to make a special snack the next school day. Berry shortcake is a tasty way to enjoy the fruits of your labor. Have the children help with all the preparation, mixing, etc.

You will need

- oven
- aprons
- mixing bowls and spoons
- measuring utensils
- cake pan—a 9 inch round or 8 x 11 inch oblong
- 3/4 cup oil, plus some for oiling the pan
- 1/2 - 3/4 cup honey, plus a little extra
- 1 1/4 cup yogurt or milk
- 2 teaspoons vanilla
- 2 1/2 cups whole wheat pastry flour
- 1 tablespoon baking soda
- 3/4 teaspoon salt
- fresh berries, enough for 2-3 tablespoons per serving
- bowls and spoons
- optional—whipped cream

What to do

1. Preheat oven to 350 degrees.

2. Mix the wet ingredients—oil, honey, yogurt or milk, vanilla—in a bowl.

3. Mix the dry ingredients—flour, baking soda, salt—in a separate bowl.

4. Combine the wet and dry ingredients.

5. Pour the batter into an oiled 9 inch round or 8 x 11 inch oblong cake pan.

6. Bake for approximately 45 minutes, until the top springs back when touched gently with your finger.

7. While the cake is baking, have the children help you rinse and slice the berries (or use them whole) and toss them with a bit of honey. When the cake is cool, spread the berries on the top of the cake, garnish with whipped cream and enjoy!

Berry Cobbler

And try some cobbler, too!

You will need

- oven
- aprons
- mixing bowls and spoons
- knives
- measuring utensils
- 8 x 8 inch baking pan
- saucepan and heat source or an electric frying pan
- 2 cups whole wheat flour, plus 1 tablespoon
- 3 teaspoons baking powder
- 1/2 teaspoon salt
- 1/3 cup butter, plus extra for greasing the pan
- 1 egg
- 1/2 cup milk
- 1 tablespoon honey or sucanat, plus 1/2 cup
- 3 cups berries
- bowls and spoons

What to do

1. Preheat oven to 425 degrees.

2. In a large bowl, mix together flour, baking powder and salt.

3. Cut in the butter until the mixture resembles coarse crumbs.

4. In a separate bowl, beat together the egg, milk and the tablespoon of honey or sucanat.

5. Add the liquid all at once to the flour mixture and stir until just moistened.

6. Combine the berries, 1/2 cup honey or sucanat and one tablespoon flour in a saucepan and heat, stirring to be sure it doesn't stick. You may need to add a little water.

7. Bring to a boil, turn down and stir until thickened a bit. Watch carefully so that it doesn't stick and burn.

8. Place the berry mixture in the bottom of a greased 8 x 8 inch pan.

9. Spoon the biscuit dough over the top.

10. Bake in a 425-degree oven for approximately 1/2 hour.

11. Let cool, eat and enjoy!

 SUPPLYING THE MISSING LINKS: BERRIES

Berry Smoothies

A very refreshing way to enjoy your berry harvest. Smoothies are a delicious and nourishing snack drink. Once you've learned to make them, you can vary the ingredients endlessly and try new fruit combinations.

You will need

- a blender
- smoothie ingredients (use organic ingredients whenever possible)—apple juice, yogurt, honey, rinsed berries. Bananas are optional, but they add flavor and thickness.
- measuring utensils
- cups

What to do

1. Use apple juice as the base. Place two cups in the blender.

2. Add 1/2 cup plain yogurt, one heaping tablespoon of honey, one cup of rinsed berries and one ripe banana.

3. Blend until smooth.

4. Taste a bit and adjust the ingredients, more honey to add sweetness, more yogurt for a more tart taste, more berries for a fruitier taste, more juice to "stretch" the smoothie so there will be enough to go around.

5. Pour into cups and drink for snack. Makes 3—4 cups.

NOTE: In hot weather, add a couple of ice cubes to make it really chilled.

Berry Preserves

Making berry preserves from your remaining harvest is not as hard as you might think, especially if you can keep the jam refrigerated until you are ready to use it. You just need a safe place to cook it for a while. If you're lucky enough to have a kitchen or hot plate in your room, you'll be rewarded with delightful fruity smells.

You will need

- colander or strainer
- berries
- large heavy pot (large enough to hold the amount of berries you'll be cooking)
- potato masher
- oven or hot plate
- long-handled wooden mixing spoon
- sweetener—preferably honey or sucanat (organic, dried juice of sugar cane that isn't refined), but you can use regular sugar.
- measuring cups
- spoons
- clean glass jars with tight-fitting lids

What to do

1. Gently rinse the berries.

2. Place them in the pot and mash them a bit with the potato masher.

3. Begin to cook them, stirring frequently, and bring the mixture to a rolling boil (bubbling all over the top).

Add a small amount of water, if necessary, to keep it from sticking. Then turn the heat down to low.

4. Stir in the sweetener 1/2 cup at a time, tasting (with a clean spoon) for sweetness. If the berries were very sweet, you won't need as much. Remember, the jam doesn't have to be as sweet as "store bought." It's nice to taste the berries.

NOTE: I usually use natural, unrefined sweeteners such as honey, maple syrup or sucanat when cooking or baking as they contain vitamins and minerals and as such are more healthful than processed, refined sugar.

5. Stir carefully and frequently during the cooking process so the jam doesn't stick and burn.

6. When it has cooked down and thickened a good bit, turn it off and let it cool. It will continue to thicken as it cools.

7. Spoon into clean glass jars and keep refrigerated until you are ready to use it.

Freezing Berries

Freezing berries is kind of the opposite of making jam, using cold instead of heat to preserve the fruit. But it is an easy way to save the whole berries for future use, and all you need is a freezer. Younger children can certainly help with this; just expect as much eating as helping!

You will need

- colander or strainer
- berries
- clean towels
- cookie sheets
- freezer space
- plastic containers with tight fitting lids (re-use those yogurt containers from the smoothies)
- adhesive tape

What to do

1. Rinse the berries carefully and lay them out on clean towels to drain.

2. Place them on the cookie sheets side by side, but keep each berry separate so cool air can circulate around them. Put them in the freezer overnight.

3. The next day remove the frozen berries from the cookie sheet and place them in a plastic container. Use adhesive tape to mark the date and the kind of berry. Just like frozen grapes, frozen berries make a very cooling and healthful summer snack.

NOTE: This is an easy way to preserve almost any fruit you'd like to save. You can even save peach slices if you first toss them with a bit of lemon juice to prevent darkening. The frozen fruits make a wonderful garnish for a birthday cake or add a special summer memory to a fruit salad.

Basket Weaving

This activity is a good challenge for the older children. But it is also wonderful for the younger children to watch and experience the adults around them doing "real" work—work that results in things that they can see and use. It is especially nice to work on the basket outdoors during playtime. The children will be very interested in what you're doing and will probably ask you if they can help.

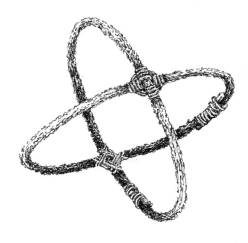

You will need

- weaving materials—you can try almost anything that feels flexible and sturdy enough to weave: willow wands—the long supple branches from the willow tree—vines, e.g., kudzu, Virginia creeper, wild grapes, honey suckle, ivy, etc. Vines work well because you can get long continuous pieces to weave. You can also, if necessary, use store bought cane for weaving, or use cane for the base and natural materials for the weaving.

- brown paper bag

What to do

1. Gather your weaving materials, strip off the leaves (the children love to help with this), coil them loosely and place them in brown paper grocery bags to "cure" for a few days. The curing can be omitted if you are pressed for time, but it helps make a better basket.

2. Select the thickest material or store bought cane. From these make two rings approximately 7-8 inches in diameter. The size of the rings will determine the size of your basket, so don't make them too big or too small.

Seven to eight inches in diameter is a good size for a first effort.

3. Place the two rings together at right angles to each other. Lash them together at the intersecting points using the "God's Eye" wrap.

Wrap the weaving material over the top post, bring it around the post and back over the post again.

Bring it to the next post, wrap the weaving material around the post and back over the post again, etc.

This makes the frame of the basket. One half-circle is the handle and the other three half-circles form the basket.

4. Now make the warp. Wrap the bottom of the basket with weaving material by running it from side to side to form the warp on the bottom of the basket. Wrap around the top edge, go down and over or under the bottom seam, and up around the opposite edge. These don't need to be tight to-gether. In fact, you can space them 1/4 - 1/2 inch apart because you need room to weave between them. Continue until the bottom of the basket is full.

5. Now comes the real weaving. Take weaving material, and starting on the sides near a God's Eye, begin to weave over and under the warp pieces. Overlap loose ends (these can be tucked in later) and continue weaving until the basket is complete. The more tightly you weave, the sturdier the basket.

6. If the children want to help, let them do some of the over and under weaving, helping them as needed.

7. Use the finished basket in your classroom to hold toys or supplies or to carry flowers or green beans in from the garden. It is wonderful for the children to experience the making of something from— apparently—nothing.

Mail Order Companies

The Ark
4245 Crestline Avenue
Fair Oaks, CA 95628
1-800-872-0064
- sets of natural building blocks
- art supplies, including modeling beeswax and crayons
- quality wooden toys and books

The Arts and Crafts Materials Institute
715 Boyleston Street
Boston, MA 02116
1-617-266-6800
- information on non-toxic art supplies

Bartlettyarns
P.O. Box 36
Harmony, ME 04942-0036
1-207-683-2251
- wool yarn

Basketville
Main Street
P.O. Box 710
Putney, VT 05346-0710
1-802-387-5509
- small wooden butter churns
- It sells wholesale. Ask them about a re-
  tail outlet near you.

Brookstone/Hard to Find Tools
5 Vase Farm Road
Peterborough, NH 03460-0803
1-603-924-9541
- composters
- all kinds of tools and gadgets

Central Shippee, Inc.
P.O. Box 135
Bloomingdale, NJ 07403
- manufacturer of wool and wool blend
  felt

Chaselle, Inc.
9645 Gerwig Lane
Columbia, MD 21046
1-301-381-9611
- colored tissue paper and art supplies

A Child's Dream
P.O. Box 2203
Shingle Springs, CA 95682
- wool felt
- wool fleece
- wool yarn
- wooden toys

Coop America
2100 M Street, NW, Suite 403
P.O. Box 18217
Washington, DC 20036
1-202-223-1881
- ecologically sound and socially responsible products of all kinds, including composters

Earth Guild
37 Haywood Street
Asheville, NC 28801
1-800-327-8448
- supplies for spinning and weaving
- natural fabric dyes

EcoDesign Company
1365 Rufina Circle
Santa Fe, NM 87501
1-505-438-3448
- natural art supplies and household products

Edible Landscaping
P.O. Box 77
Afton, VA 22920
1-804-361-9134
- catalog of edible landscaping plants

Hans Schumm Woodworks
R.D. 2, Box 233
Ghent, NY 12075
1-518-672-4685
- wooden farms, villages, trucks, etc.

HearthSong
P.O. Box B
Sebastopol, CA 95473
1-800-325-2502
- sets of natural building blocks
- beeswax for candle dipping
- beeswax sheets for rolling candles
- modeling beeswax
- candle decorating beeswax
- beeswax crayons
- Stockmar watercolor paints
- also many wooden toys and games

Heartwood Arts
R.D. 1, Box 126
Route 44/45
Modena, NY 12548
1-914-883-5145
- wooden houses, castles, people

Meadowbrook Herb Garden
Route 138
Wyoming, RI 02898
1-401-539-7603
- modeling beeswax
- wooden toys

Mountain Sunrise
c/0 Peg Smith
279 Swanzey Lake Road
Winchester, NH 03470
1-603-357-9622
- plant-dyed wool fleece, 100% wool felt and finger knitting yarn

Nova
27 Eagle Street
Spring Valley, NY 10977
1-914-426-3757
- sets of natural building blocks
- 100% wool yarn
- wool felt
- modeling beeswax
- beeswax sheets for rolling candles
- candle decorating beeswax
- wool felt
- beeswax crayons
- Stockmar watercolors
- wooden toys

Plow and Hearth
301 Madison Road
P.O. Box 830
Orange, VA 22960-0492
1-800-627-1712
- composters
- garden tools
- bird baths and feeders

Real Goods
966 Mazzoni Street
Ukiah, CA 95482
1-800-762-7325
- an interesting catalog with all kinds of products to promote energy independence in one way or another, including composters

River Farm
Route 1, Box 401
Timberville, VA 22853
1-800-USA-WOOL
- wool fleece and roving

Seventh Generation
Colchester, VT 05446-1672
1-800-456-1177
- a catalog full of environmentally friendly products of all kinds, including composters

Smith and Hawken
25 Corte Madera
Mill Valley, CA 94941
1-415-383-2000
- composters
- seeds
- garden tools
- excellent children's tools

Solutions
P.O. Box 6878
Portland, OR 97228-6878
1-800-342-9988
- composters

Strauss and Company
1701 Inverness Avenue
Baltimore, MD 21230
1-800-638-5555
- cheesecloth wholesale

Textile Reproduction
c/o Edmund and Kathleen Smith
Box 48
West Chesterfield, MA 01084
1-413-296-4437
- plant dyed wool
- wool yarn
- wool felt

Vermont Country Store
P.O. Box 3000
Manchester Center, VT 05255
1-802-362-4647
- oil cloth
- lots of "old-fashioned" things

Vidar Goods
P.O. Box 808
College Park, MD 20740
1-301-982-2511
- wooden animals

Walnut Acres
Penns Creek, PA 17862
1-800-433-3998
- organically grown wheat berries (kernels)
- quality organically grown foods of all kinds, from soup to nuts!

West Earl Woolen Mill
R.D. 2
Ephrata, PA 17522
1-717-859-2241
- wool batting

Yolo Wool Products
Route 3, Box 171-D4
Woodland, CA 95695
- wool batting for stuffing
- wool knitting yarn

Organizations

Alliance for Environmental Education
10751 Ambassador Dr. Suite 201
Manassas, Virginia 22110
1-703-335-1025

American Forestry Association
P.O.Box 2000
Washington, D.C. 20013
1-202-667-3300
- Project: Global Releaf: campaign to encourage planting and caring for trees to improve the environment.

Association of Waldorf Schools of
  North America
c/o David Alsop
3750 Bannister Rd.
Fair Oaks, California 95628
- Information on Waldorf Education, teacher training centers, list of Waldorf schools.

Center for Environmental Information
46 Prince St.
Rochester, New York 14607
1-716-271-3550
- Accurate and comprehensive information on environmental issues. Publications, educational programs and information services.

National Arbor Day Foundation
211 N. 12th St., Suite 501
Lincoln, Nebraska 68508
1-402-474-5655
- Educational organization dedicated to tree planting and conservation.

National Audubon Society
950 Third Ave.
New York, New York 10022
1-212-832-3200
- Dedicated to protecting wildlife and its habitats.

National Wildlife Federation
1400 16th St. N.W.
Washington, D.C. 20032
1-202-797-6800
- Distributes periodicals and educational materials; sponsors outdoor education programs in conservation.

Waldorf Kindergarten Association of
  North America
9500 Brunett Ave.
Silver Spring, Maryland 20901
1-301-565-2282
- Information on Waldorf Kindergarten Education; publishes twice-yearly newsletter; booklists.

Background Reading for Teachers

Periodicals/Newsletters

Buzzworm: The Environmental Journal, Subscriptions: P.O.Box 6853, Syracuse, New York 13217, 1-800-825-0061.

E—The Environmental Magazine, Subscriptions: P.O. Box 6667, Syracuse, New York 13217. 1-800-825-0061

The Earthwise Consumer Newsletter, P.O. Box 279, Dept. NNE, Forest Knolls, California 94933, 1-415-488-4614. Debra Lynn Dadd's newsletter to keep you up to date on the best "environmentally friendly" products as well as environmental issues that relate to consumer choices.

Garbage: The Practical Journal for the Environment, Subscriptions: P.O. Box 51647, Boulder, Colorado 80321-1647, 1-800-888-9070

The Environment

Blueprint For A Green Planet: Your Practical Guide To Restoring The World's Environment by John Seymour and Herbert Girardet. New York: Prentice Hall, 1987. Clear, well-illustrated descriptions of different environmental problems with simple, excellent suggestions for positive action. Some of the information may need updating.

Clean & Green: The Complete Guide To Nontoxic and Environmentally-Safe Housekeeping by Annie Berthold-Bond. Woodstock, New York: Ceres Press, 1990. 500+ recipes for natural, homemade, cleaning products.

Earth Right—Every Citizen's Guide by H. Patricia Hynes. Rocklin, California: Prima Publishing & Communications, 1990. What you can do in your home, work place and community to save our environment. Shows you how to think globally and act locally.

50 Simple Things Kids Can Do To Save The Earth by The Earth Works Group. Kansas City, Missouri: Andrews and McMeel, 1990. Good background reading for teachers.

50 Simple Things You Can Do To Save The Earth by The Earth Works Group. Berkeley, California: Earthworks Press, 1989. Concise descriptions of various environmental problems and lots of clearly written suggestions for things which you can do to make a difference from stopping the flow of junk mail to starting a compost pile.

The Green Consumer by John Elkington, Julia Hailes and Joel Makower. New York: Penguin Books, 1990. Discussion of green consumerism and how your everyday purchases can affect the Earth. Comprehensive product guide. Excellent resource guide to books, organizations and agencies.

The Green Consumer Supermarket Guide: Brand Name Products That Don't Cost The Earth by Joel Makower. New York: Penguin Books, 1991. Names names!

The Green Lifestyle Handbook by Jeremy Rifkin, editor. New York: Henry Holt & Co., 1990. 1001 ways we can heal the Earth by choosing a lifestyle that will help create a sustainable future.

The Green Pages—Your Everyday Shopping Guide to Environmentally Safe Products by The Bennett Information Group. New York: Random House, 1990. Anatomy of a green product and guide to environmental issues with both a market shopper's and mail order shopper's guide that gives brand names.

The Next Step: 50 More Things You Can Do To Save The Earth by The Earth Works Group. Kansas City, Missouri: Andrews and McMeel, 1991. New suggestions about things you can do for yourself, for your neighbors and for your community as well as 16 individual stories about people who made a difference and how.

Nontoxic and Natural: How To Avoid Dangerous Everyday Products and Buy Or Make Safe Ones by Debra L. Dadd. Los Angeles: J.P. Tarcher, 1984.

The Nontoxic Home: Protecting Yourself and Your Family From Everyday Toxics and Health Hazards by Debra L. Dadd. Los Angeles: J.P. Tarcher, 1986.

Nontoxic, Natural and Earthwise: How To Protect Yourself and Your Family From Harmful Products and Live In Harmony With The Earth by Debra L. Dadd. Los Angeles: J.P. Tarcher, 1990. Comprehensive listing of healthful products available through retail and mail order sources, and a rating system that indicates both their safety and their environmental impact. Excellent resources section.

Our Earth, Ourselves by Ruth Caplan and the Staff of Environmental Action. New York: Bantam Books, 1990. The action-oriented guide to help you protect and preserve our environment. Clear discussion of environmental problems, point-by-point blueprint for action, and profiles of individuals who have made a difference. Excellent organization directory.

Save Our Planet—750 Everyday Ways You Can Help Clean Up the Earth by Diane MacEachern. New York: Dell Publishing, 1990. Organized around the areas where we live and work such as home, garden, garage, supermarket, school, office, etc. with ideas for simple changes which, while making no appreciable difference in the way we live, can have substantial positive impact on the world around us. Appendix lists books, catalogs, agencies and organizations.

The Classroom

Children At Play—Preparation For Life by Heidi Britz-Crecelius. Edinburgh: Floris Books, 1979. Excellent discussion of the nature of children's play and ways in which to support its development in both the classroom and the home.

The Children's Year by Stephanie Cooper, Christine Fynes-Clinton and Marye Rowling. Gloucestershire: Hawthorn Press, 1986. a seasonally organized book of crafts and clothes for children and parents to make.

Festivals, Family and Food by Diana Carey and Judy Large. Gloucestershire: Hawthorn Press, 1982. Wonderful source book of seasonally organized songs, stories, activities and recipes for use in home and school.

Let Us Form A Ring—An Acorn Hill Anthology edited by Nancy Foster. A wonderful little collection of songs, verses, stories and circle games organized seasonally. Available from Acorn Hill Children's Center, 9500 Brunett Avenue, Silver Spring, Maryland 20901.

The Nature Corner—Celebrating the Year's Cycle With a Seasonal Tableau by M v Leeuwen and J. Moeskops. Edinburgh: Floris Books, 1990. Detailed descriptions with lovely color illustrations of how to create a seasonal garden or nature corner. Includes patterns and directions for making little people and accessories.

Spinning and Dyeing the Natural Way by Ruth Castino. New York: Van Nostrand Reinhold Company, 1974. Well-illustrated guide to all aspects of working with wool. Shows you how to wash, dye, card and spin wool. You, too, can learn how to use a drop spindle!

Toymaking With Children by Freya Jaffke. Edinburgh: Floris Books, 1990. Excellent instructions for creating toys from natural materials for and with children. Discussion of the nature of play and detailed descriptions and patterns for making various building materials, dolls, knitted animals, simple string puppets and much more.

The Outdoors

The Art of Composting. Available free from The Metropolitan Service District, 2000 S.W. 1st Ave., Portland OR 97201.

Backyard Bird Feeding. How to attract different species with the right food and type of feeder. Advice on placement and cleaning. 24 pp. Pamphlet number 558X available free from U.S. General Services Administration. Send $1.00 to cover handling to S.James, Consumer Information Center-W, P.O. Box 100, Pueblo, Colorado 81002.

Garden Birds: How To Attract Birds To Your Garden by Dr. Noble Proctor. Emmaus, Pennsylvania: Rodale Press, 1986. Contains a wonderful directory of garden birds with color illustrations and discussions of how to garden for birds, food and feeders, and nest boxes.

Homes For Birds. How to choose the right house to attract the species you want. Describes the characteristics of various birds, nesting behaviors, and how to protect the houses against predators. 32 pp. Pamphlet number 572X available free from U.S. General Services Administration. Send $1.00 to cover handling to S.James, Consumer Information Center-W, P.O. Box 100, Pueblo, Colorado 81002.

Let It Rot! The Home Gardener's Guide to Composting by Stu Campbell. Pownal, Vermont: Garden Way Publishing, 1975. An easy-to-follow excellent reference.

The Organic Gardener by Catharine Osgood Foster. New York: Random House, 1972. Simple, well-written description of how to do garden organically from start to finish. The book that got me started.

The Rodale Guide to Composting by Jerry Minnick, Marjorie Hunt, et al. Emmaus, PA: Rodale Press, 1979. An encyclopedia of composting.

Sharing Nature With Children by Joseph Cornell. Nevada City, California: Dawn Publications, 1979. The classic nature awareness guidebook for parents and teachers. Excellent background material for teachers who want to develop in themselves a new sensitivity toward nature.

The Rodale Press, Inc., 33 East Minor Street, Emmaus, Pennsylvania 18049 is an excellent source for myriad books on all aspects of gardening organically, composting, etc.

Picture Books for Children

Note: When choosing picture books for use with young children, be particularly aware of the quality of the illustrations - they should be clear, colorful and, especially, beautiful, as they are representing the world to the children. Also, never forget the power of simply telling stories to the children. No picture can ever be more beautiful or more powerful than the one they create themselves in their own mind's eye!

Fall

Apples and Pumpkins by Anne Rockwell, illustrated by Lizzy Rockwell New York: Macmillan Publishing Co., 1989. Simple story of taking an autumn trip to a farm to get pumpkins and apples. Pleasant illustrations (except for the scary masks in the last picture!).

Around The Year by Elsa Beskow. Edinburgh: Floris Books, 1988. Lovely verses and pictures for each month of the year.

Chipmunk Song by Joanne Ryder, illustrated by Lynne Cherry. New York: E.P. Dutton, 1987. Imagine you are a little chipmunk and go about all the activities necessary to prepare for winter. Very detailed illustrations.

From Seed To Pear by Ali Mitgutsch. Minneapolis: Carolrhoda Books, Inc., 1981. One of a long series of what Carolrhoda Books call their start-to-finish books. This one tells simply the story of a pear tree and how it grows from a seed. This press's other books tell start-to-finish stories of many familiar things such as shoes, bricks, honey, bread, butter and furniture, just to name a few.

Johnny Appleseed by Reeve Lindbergh, illustrated by Kathy Jakobsen. Boston: Little, Brown & Co., 1990. Lovely, detailed folk-art paintings illustrate the poem-story of Johnny Appleseed.

My Favorite Time of Year by Susan Pearson, illustrated by John Wallner. New York: Harper & Row Publishers, 1988. Starting with fall, a family enjoys the special activities that each season of the year brings, with nice illustrations throughout the book.

Night In The Country by Cynthia Rylant, illustrated by Mary Szilagyi. New York: Macmillan Publishing Co., 1986. Colorful illustrations and simple text describe the sights and sounds of a night in the country.

Once There Was A Tree by Natalia Romanova, illustrated by Gennady Spirin. New York: Dial Books, 1985. Originally published in Russia, the story of an old tree stump and the interdependence of all the creatures who are attracted to it. There is a lovely, old-world quality to the illustrations.

The Ox Cart Man by Donald Hall, illustrated by Barbara Cooney. New York: Viking Press, 1979. Describes the day-to-day life throughout the changing seasons of an early 19th century New England family and how they had to make much of what they used to live, with country primitive style illustrations.

The Pumpkin Blanket by Deborah Turney Zagwyn. Berkeley, California: Celestial Arts, 1990. Lovely watercolors illustrate the story of a little girl who sacrifices her beloved blanket to save the pumpkins in the garden from the frost. She sometimes plays in the compost pile!

Pumpkin Pumpkin by Jeanne Titherington. New York: Scholastic Inc., 1989. Beautifully illustrated story of a little boy who plants pumpkin seeds and, eventually, carves his own jack-o'-lantern.

The Seasons Of Arnold's Apple Tree by Gail Gibbons. New York: Harcourt Brace Jovanovich, 1984. As the seasons pass, Arnold enjoys a variety of activities as a result of his apple tree. Includes a recipe for apple pie and a description of how an apple cider press works. Bright, child-like drawings.

Squirrels by Brian Wildsmith. Oxford: Oxford University Press, 1974. Large, colorful illustrations tell the story of the squirrel and his habits, an animal even city children know.

The Year At Maple Hill Farm by Alice and Martin Provensen. New York: Atheneum, 1978. A book about the seasonal changes on a farm and in the surrounding countryside, with lovely, gentle yet detailed illustrations.

Winter

Animals Of The Night by Merry Banks, illustrated by Ronald Himler. New York: Charles Scribner's Sons, 1990. Describes in simple text and lovely pictures the various animals that are active only at night.

The First Snowfall by Anne and Harlow Rockwell. New York: Macmillan Publishing Co., 1987. Simple text and pictures tell the story of what a child does on a snowy day.

Keep Looking by Millicent Selsam and Joyce Hunt, illustrated by Normand Chartier. New York: Macmillan Publishing Co., 1989. Very detailed and beautiful illustrations with simple, direct text tell the story of all the animals you will see near an old house in winter if only you keep looking.

The Midnight Farm by Reeve Lindberg, illustrated by Susan Jeffers. New York: Dial Books, 1987. The secrets of the dark are revealed in this rhythmic poem describing a farm at midnight, with beautiful illustrations.

Ollie's Ski Trip by Elsa Beskow. Edinburgh: Floris Books, 1989. Ollie meets Jack Frost, King Winter, Mrs. Thaw and, eventually, Spring on a trip into the forest; beautifully illustrated.

Owl Moon by Jane Yolen, illustrated by John Schoenherr. New York: Philomel Books, 1987. On a winter's night, under a full moon, a father and daughter trek into the woods to see the Great Horned Owl. Illustrated with lovely, wintry watercolor paintings.

Stopping By Woods On A Snowy Evening by Robert Frost, illustrated by Susan Jeffers. New York: Holt, Rinehart and Winston, 1969. Lovely, soft, mostly black and white illustrations for each line of the famous poem.

The Tomten by Astrid Lindgren, illustrated by Harald Wiberg. New York: Coward-McCann, Inc., 1961. The Tomten, a gnome-like creature, visits the sleeping inhabitants of a farm one winter night, with soft watercolor illustrations.

Under Your Feet by Joanne Ryder, illustrated by Dennis Nolan. New York: Four Winds Press, 1990. Lovely, soft illustrations give a glimpse into the lives of animals we don't always see during different seasons of the year.

Winter Magic by Eveline Hasler, illustrated by Michele Lemieux. New York: William Morrow & Co., 1984. Fanciful tale in which Peter's cat, Sebastian, takes him out into a snow-covered world to show him the secrets of winter, with softly brushed paintings.

Spring

Aranea: A Story About A Spider by Jenny Wagner, illustrated by Ron Brooks. Scarsdale, New York: Bradbury Press, 1975. Soft, black and white drawings illustrate the story of a spider who, despite interruptions, spends her days and nights spinning perfect webs.

Baby Animals by Margaret Wise Brown, illustrated by Susan Jeffers. New York: Random House, 1989. Tells the story of a day—morning, noon and evening—in the life of a little girl and the baby animals who live on her family's farm, with soft, detailed illustrations.

Bringing The Rain To Kapiti Plain by Verna Aardema, illustrated by Beatriz Vidal. New York: Dial Press. Rhythmic African tale of how the rains came that shows the interdependence of nature's kingdoms.

Gilberto and the Wind by Marie Hall Ets. New York: Viking Press, 1963. Story of a little boy and his adventures with the wind.

Home In The Sky by Jeannie Baker. New York: Greenwillow Books, 1984. A story of how birds (pigeons) are cared for even in the city, with interesting collage illustrations.

The Listening Walk by Paul Showers, illustrated by Aliki. New York: HarperCollins Publishers, 1991. Gentle watercolors illustrate the simple story of a child who takes a walk with her father but doesn't talk. It's a listening walk!

Miss Rumphius by Barbara Cooney. New York: Viking Press, 1982. Classic tale of a little girl who loved the sea and grew up wanting to make the world a more beautiful place which she does by planting flowers.

My Cousin Katie by Michael Garland. New York: Thomas Y. Crowell, 1989. Story of all the things Katie does on the farm from collecting eggs and picking apples to helping her father fix the tractor, with interesting, realistic illustrations.

My Spring Robin by Anne Rockwell, illustrated by Harlow and Lizzy Rockwell. New York: Macmillan Publishing Co., 1989. A little girl notices lots of signs of spring while she waits for the return of her friend, the robin, with simple, colorful illustrations.

Pelle's New Suit by Elsa Beskow. Edinburgh: Floris Books, 1989. The story of how a little boy gets a new suit from his lamb. Shows how the wool is sheared, carded, spun, dyed and woven.

The Story of the Root Children by Sibylle von Olfers. Edinburgh: Floris Books, 1990. Enchanting story of the root children who are awakened by Mother Earth in the spring and all they do till she calls them home again in the fall.

This Year's Garden by Cynthia Rylant, illustrated by Mary Szilagyi. New York: Macmillan Publishing Co., 1987. Warm, colorful pictures and simple text tell the story of a family's garden from planning through harvest.

A Tree Is Nice by Janice May Udry, illustrated by Marc Simont. New York: HarperCollins Publishers, Harper Trophy edition, 1987. Tells all the wonderful reasons why trees are nice.

Summer

The Flowers' Festival by Elsa Beskow. Edinburgh: Floris Books, 1991. Fanciful story of Lisa who joins the flowers for their Midsummer festival.

A House Of Leaves by Kiyoshi Soya, illustrated by Akiko Hayashi. New York: Philomel Books, 1987. Gentle watercolors illustrate the story of a little girl who creeps under a bush to get out of the rain and finds new insect friends there.

Is This A House For Hermit Crab? by Megan McDonald, illustrated by D.D. Schindler. New York: Orchard Books, 1990. Gentle, pastel drawings illustrate the story of hermit crab who outgrows his old house and ventures out to find a new one.

Peter's Adventures In Blueberry Land by Elsa Beskow. Edinburgh: Floris Books, 1989. Peter has a magical adventure when he goes in search of berries for his mother's birthday.

A Potter by Douglas Florian. New York: Greenwillow Books, 1991. Simple, colorful illustrations and simple text tell the whole story of what a potter does with clay.

Rain Talk by Mary Serfazo, illustrated by Keiko Narahashi. New York: Margaret K. McElderry Books, 1990. A little girl experiences a wonderful summer rain and especially all the different sounds it makes, with lovely illustrations.

Salt Hands by Jane Chelsea Aragon, illustrated by Ted Rand. New York: E.P. Dutton, 1989. In the middle of the night, a young girl wakens and goes outside to let a deer lick salt from her hand. Lovely, soft illustrations.

The Turtle and the Moon by Charles Turner, illustrated by Melissa Bay Mathis. New York: Dutton Children's Books, 1991. Beautifully illustrated and fanciful story of a turtle who learns to play with the moon.

Where Butterflies Grow by Joanne Ryder, illustrated by Lynne Cherry. New York: Dutton Children's Books, 1989. Beautifully illustrated story of how a butterfly comes to be.

Where The River Begins by Thomas Locker. New York: Dial Books, 1984. Two young boys and their grandfather go on a camping trip to find the source of the river that flows past their home, with extraordinarily beautiful landscape paintings.

Wild Wild Sunflower Child Anna by Nancy White Carlstrom, illustrated by Jerry Pinkney. New York: Macmillan Publishing Co., 1987. Beautiful illustrations and poetic text tell the story of a little girl who plays outdoors on a summer day.

Index